SPICES

Volume II

Morphology · Histology · Chemistry

by

John W. Parry

CHEMICAL PUBLISHING COMPANY, INC · NEW YORK · 1969

Chemical Publishing Company, Inc.
New York

Printed in the United States of America

Dedicated to my wife
MARY PARRY

PREFACE TO THE SECOND EDITION

The publication of a new edition of this book has afforded an opportunity to add chapters on chervil and tarragon, two herbs that have enjoyed a favored place in European kitchens since ancient Greek and Roman times and that have found increased consumer demand in America in recent years.

It has also afforded an opportunity to add recent information on the composition of the essential oils of the spices as reported in *Chemical Abstracts* of the American Chemical Society and in other scientific journals.

Since almost all of the demand for sweet basil is now met by American growers, the chapter and line drawings of this herb have been revised to cover the domestically cultivated product.

Other additional information has been incorporated into the text, and the glossary has been considerably expanded.

The author would like to take this opportunity to extend his thanks to Sr. Miguel Pascual Giménez, Ingeniero Director, Instituto Nacional de Investigaciones Agronómicas, Ministerio de Agricultura, Murcia, Spain, for information on Spanish paprika; to M. André Darbonne, Syndicat National des Producteurs, Ramasseurs & Collecteurs de Plantes Médicinales, Aromatiques et Industrielles, Milly-La-Forêt, France, for information on chervil and tarragon; to Mr. F. Caligari, Spice Islands Company, California, for his letter on chervil; to Mr. A. W. Pastor, San Jacinto Spice Ranch, Inc., Inglewood, California, for information on the production of sweet basil and tarragon, and for the samples of whole dried leaves of these herbs which he so kindly sent me; and to Mr. Marshall W. Neale, The American Spice Trade Association, New York, for statistical and other information, and for his assistance in obtaining the above samples.

John W. Parry

PREFACE TO THE FIRST EDITION

This work is intended as a short reference book on the structure and chemical composition of spices; to provide helpful information for those whose responsibility it is to examine samples of spices with the microscope; to provide a guide to the microscopy of spices for students of food analysis; to interest students of plant histology and others willing to make a reasonable expenditure of time and study in the structure of spices; to provide information on the structure and chemical composition of spices for the use of all those engaged in, or associated with, the production, manufacture, and distribution of spices; and to convey to the general reader some idea of the external form, the complex but interesting internal structure, and the chemistry of spices.

Part 1 is in no sense an exhaustive enquiry into the anatomy of the parts of plants used as spices, but an attempt to describe as clearly and simply as possible the most important features of the external form and internal structure of the various spices.

Part 2 deals with the chemical composition of spices. The basic chemical composition of spices is much like that of many other plants. However, their volatile oils, oleoresins, pigments, glucosides, and, in some, fixed oils, are chemical compounds of special interest. It is these that make the spices so desirable as food-flavoring agents. These substances have been a matter for investigation by a great number of people over a very long period of time, and are still the subject of enquiry by many capable chemists. Much is known, and much is yet to be learned, about the complex organic chemical compounds contained within the spices. No attempt has been made to discuss the chemistry of the numerous complex chemical compounds involved. Such information can be found in larger reference books, abstracts, and papers.

Part 3 consists of photomicrographs of the spices.

For information on the production of spices the author expresses his gratitude to the Director of Agriculture, Kuala Lampur, Malaya Union; Director of Agriculture, Department of Agriculture, Peradeniya, Ceylon; Director of Commerce, Department of Commerce, Ceylon; Director of Agriculture,

Department of Agriculture, Chepauk, Madras, India; Director of Agriculture, Department of Agriculture, Zanzibar, East Africa; Director of Agriculture, Department of Agriculture, Sierra Leone, Africa; Director General, Ministério da Economia, Lisbon, Portugal; Grenada Co-operative Nutmeg Association, Grenada, British West Indies; to Mrs. Ezra Winter, House of Herbs Inc., Salisbury, Connecticut, for samples of peppermint and spearmint; and to Mr. Marshall W. Neale, Director of Information, American Spice Trade Association, New York, for his assistance in obtaining these samples.

John W. Parry

CONTENTS

Part 1

MORPHOLOGY AND HISTOLOGY

INTRODUCTION

Morphology

Morphology is the study of external organic form. In this work, it is the study of the external form of the parts of plants employed to give flavor, relish, and piquancy to foods and found as such in the products of the spice merchant.

It must be kept in mind that spices are *dried* products. The size, shape, color, and general appearance of any one of them will vary to some extent from similar properties in their fresh, living counterparts. Should any descriptive term not be understood it will be found explained in the glossary.

Histology

By histology is meant here the microscopic structure or anatomy of the spices and a study of the cells and tissues of the parts of plants used as spices.

Such a study involves the use of a microscope and other equipment. The microscope is an important, indispensable item. Many fine types are manufactured today. One may buy a very high quality microscope equipped with an incorporated light source, aplanatic condenser, mechanical stage, apochromatic objectives, quadruple nosepiece, compensating oculars, and other advanced features.

However, such an elaborate and expensive instrument is not required for our purpose. An inexpensive microscope equipped with coarse and fine adjustment, a $\times 10$ ocular, two objectives, $\times 10$ and $\times 25$, and substage condenser, will do most of the work. A little more money will buy a very serviceable microscope equipped with a regular Huygenian $\times 10$ ocular, a revolving triple nosepiece, three achromatic objectives, coarse and fine adjustment, a revolving stage, and an Abbe condenser, which will give excellent results. The three objectives recommended are $\times 10$, $\times 25$, and $\times 50$. Very high power is not required in a study of the tissues and cells of plants.

For cutting sections a microtome is a very efficient precision instrument, but satisfactory sections can be cut with a flat-sided razor-type knife obtainable from any laboratory supply house at small expense, or with a safety razor blade. The safety razor blade is a very useful instrument, especially in obtaining sections of very small fruits or seeds.

Inexpensive polarizing apparatus can be easily and quickly made by cutting two discs from a small sheet of Polaroid, one large enough to fit in the filter ring below the substage condenser, or to be otherwise mounted in that position, to act as the polarizer; and one small enough to lie on top of the ocular, to act as the analyzer. The disc to be used as the analyzer can be fitted into a bored cap made from the end of a small, cylindrical paper or plastic box of suitable diameter. This is a useful arrangement and permits the easy rotation of the analyzer. Polarized light is a helpful aid in the study and identification of starch, crystals, and lignified tissues.

Measurements can be made with an ocular micrometer. Such a micrometer consists of a glass disc on which is engraved a scale of divisions, usually 50, without value. The disc is laid upon the diaphragm within the ocular, scale side downward. To give a value to the divisions of the ocular micrometer they must be calibrated with those of a stage micrometer. The scale of the stage micrometer is engraved upon a glass slide and usually consists of 100 divisions, each with a value of 0.01 mm. or 10 microns. The stage micrometer is brought into focus and the number of divisions covered by the entire ocular micrometer scale is noted, multiplied by 10 and divided by 50. This will give the value in microns of each division of the ocular micrometer. For example, let us say that the entire micrometer scale covers 69 divisions of the stage micrometer scale, then 69×0.01 mm equals 0.69 mm, and 0.69 mm $\times$ 1000 equals 690 microns (or simply 69×10 equals 690), and 690 microns divided by 50 equals 13.8 microns to each division of the ocular micrometer scale.

Calibrations must be made for each combination of ocular and objective. The higher the power the less the number of microns to each division of the ocular micrometer scale. Glass ocular and stage micrometers are inexpensive and obtainable from any laboratory supply house.

Dissecting needles can be obtained from a laboratory supply house, or they can be made by attaching ordinary sewing or darning needles to wooden handles. A bent needle can be obtained by heating a sewing or darning needle in a flame until it is red hot, bending, reheating, and tempering in cold water. Dissecting needles are useful for pulling apart fibrous tissues.

A pair of fine-pointed forceps will be required, and a pair of small scissors should be at hand.

A number of glass slides and cover glasses will be required; square cover glasses are recommended. Water can be contained in any small table glass. Medicine droppers are excellent for conveying liquids to slides.

A spice may be soaked in water contained in a small glass, but if it requires boiling then a pyrex beaker must be used. A beaker of pyrex glass may be safely employed over a bunsen burner, on a hot plate, or on a kitchen stove; a useful size is 50 ml.

A supply of household bleach, concentrated sulfuric acid, iodine-potassium iodide solution, and chlor-zinc-iodine will suffice for reagents and stains. If desired, a supply of phloroglucinol and hydrochloric acid may be included. Phloroglucinol ((5% in 90% alcohol) followed by hydrochloric acid diluted with an equal volume of water, is a useful stain for confirming the presence of lignified tissue. Lignified cell walls stain red or reddish-violet with phloroglucinol.

A hand magnifying glass is useful, or better still, the type of magnifier used by watchmakers. This is attached to a narrow, spring-like metal band which encircles the head and holds the lens in front of the eye. With its use both hands are free to section or dissect the spice.

It is the ground spice that is usually examined under the microscope. But before the analyst can identify a spice under examination he must be familiar with the tissues characteristic of that spice. A study of sections of whole spices will enable the analyst to identify the fragments of tissues of ground spices seen in the microscope field. Failing the knowledge obtained from such a study he must have reference to this or other texts on the microscopical structure of spices.

It is nearly always necessary to soak the spice in water to soften it sufficiently for sectioning. Some spices are very hard and must be softened by boiling in water; a little sodium hydroxide or a pinch or two of washing soda (sodium carbonate) added to the water will assist the process.

After the spice is soft enough to section, it should be held firmly between the forefinger and thumb, or held firmly against the surface of a glass slide laid upon the work table, and the desired section cut with a cutting knife or razor blade. If the spice cannot be held firmly with the hand it may be embedded in paraffin wax. This will provide support and leave no foreign material clinging to the section. If the wax does not fall away cleanly it can be removed by immersing the section in hot water. A transverse (cross) section is obtained by cutting across the longitudinal axis, a longitudinal section by cutting through the longitudinal axis, and a tangential section by cutting parallel to the longitudinal axis but not through the center of the spice.

The section is placed in a drop or two of water on a glass slide and covered with a cover glass. If the edge of the cover glass is brought into contact with the water and the glass lowered gently with the aid of a dissecting needle, no air bubbles will be left between the cover glass and the slide. The section may then be examined with the low power of the microscope. Higher powers can be used to permit a study of the cells, cell inclusions, and tissues in greater detail.

Seen at this time, the cells will be partly expanded as a result of the soaking process, and the cell inclusions will be visible. The pigments may be decolored and the cells further expanded by treatment with dilute household bleach. Sections may be bleached on the slide by drawing the bleaching solution under the cover glass. This is done by applying a piece of absorbent paper tissue or blotting paper to one edge of the cover glass and the bleaching reagent to the other. The water is drawn off and replaced by the bleach. Many bubbles arise in the process of bleaching but most of these are removed when the bleach is replaced with water after the section is decolored. The few bubbles remaining will not interfere with the study of the subject. If it is now desired to stain the section, chlor-zinc-iodine is drawn under the cover glass. Cuticle, cutinized, and lignified cell walls will stain yellow; cellulose walls, blue; starch, dark blue; proteins, brown. Regarding lignified cells it should be said that if sections are left too long in the bleaching solution, all or part of the lignin may be dissolved and the walls will give a blue color with chlor-zinc-iodine (cellulose).

Unbleached sections may be stained by treatment with iodine-potassium iodide solution followed by chlor-zinc-iodine. Color reactions will be similar to those given above except for such interference as pigments present. The presence of lignified tissues may be confirmed by staining another section with phloroglucinol.

Iodine-potassium iodide solution stains starch dark blue; aleurone, yellow to brown. It should be remembered that boiling, and solutions of alkali hydroxides, cause swelling and destruction of starch granules. For a microscopic study of starch the spice should be softened in cold or warm water, sectioned, and examined in water mount without further treatment. The presence of starch can be confirmed by drawing iodine-potassium iodide solution under the cover glass.

Concentrated sulfuric acid dissolves cellulose quickly; chars and blackens lignin, but does not alter cuticle or cutinized cell walls. Suberized cell walls are also unaffected by concentrated sulphuric acid.

Ground spices may be mounted directly in a drop or two of water placed on a glass slide, overlaid with a cover glass, and examined under the microscope. Ground spices are usually of requisite fineness for direct examination but if the fragments appear too coarse or too densely packed they may be rubbed out by applying gentle pressure to the cover glass and making a to-and-fro motion. A pencil with a clean eraser on the end is useful for this purpose. There is enough friction between eraser and glass to permit the movement of the cover glass to-and-fro without messing it up as occurs when using the finger, necessitating the removal, cleaning, and replacing of the cover glass. Bleaching and staining, if required, may be carried out in the manner described above.

Tissues will now be discussed. The simple tissues are parenchyma, collenchyma, and sclerenchyma. Parenchyma tissue is the most common and consists of cells with thin or moderately thick cellulose walls, with or without intercellular spaces. Parenchyma cells in living plants are approximately isodiametric, but in the dried spices they are nearly always collapsed and compressed, and even when expanded with the aid of reagents may fall short of their original shape, frequently appearing irregular or elongated. Parenchyma cells occur in all of the spices; they are abundant in the mesocarp of fruits, the mesophyll of leaves, and in cortex tissue. Epidermal cells are modified parenchmya cells, frequently polygonal in shape in surface view, rectangular or oblong-rectangular in cross section.

Collenchyma is a supporting tissue, and consists of cells with irregularly thickened walls or walls thickened at the corners of the cells. The walls are composed of cellulose and are highly refractive, appearing bright and shiny; they stain blue with clor-zinc-iodine. Collenchyma occurs in the pedicels of the spice fruits of the Umbelliferae except cumin; in the leaves, petioles, and stems of the spice herbs of the Labiatae; and in some other spices.

Sclerenchyma is a supporting, strengthening and protective tissue consisting of cells with thick, lignified walls. Sclerenchyma occurs in many forms. The chief kinds of sclerenchyma cells are stone cells and sclerenchyma fibers. Stone cells vary greatly in size and shape, but are not much longer than broad and have very thick, pitted walls. Sclerenchyma fibers are elongated cells, usually with pointed or tapering ends, and with thick, usually pitted, walls. Sclerenchyma is present to some extent in nearly all of the spices.

Complex tissues are phloem and xylem. They form the conducting or vascular tissue of the plant. In the widest sense, phloem includes sieve tubes, companion cells, parenchyma, and fibers. The parenchyma is known as phloem parenchyma, and the fibers as phloem or bast fibers. Xylem includes vessels, parenchyma, and fibers. The parenchyma is known as xylem parenchyma, and the fibers as xylem or wood fibers. Fibers are not always present in phloem and xylem. In the text, the term *vascular bundle* is used to indicate a bundle of phloem and xylem without fibers, and the older term *fibrovascular bundle* to indicate a bundle of phloem and xylem with fibers. A vessel is a tubelike series of cells characterized by lignified thickening deposited in various ways on the inside of the walls. The degree of lignification varies. Vessels are known as annular, spiral, scalariform, reticulated, or pitted. In annular vessels the thickening is laid down in rings; in spiral vessels in the form of spiral coils; in scalariform in transverse bars; in reticulated vessels in the form of an irregular network; in pitted vessels the thickening is nearly uniform except for small areas or pits.

The drawings and photomicrographs appearing in the text are by the author. Space does not allow all the descriptive drawings that could be made, nor the inclusion of all the photomicrographs which are on hand, to illustrate the histological structure of the spices, but it is hoped that those that appear in this work will be of assistance to the analyst, investigator, or student, and of interest to the general reader.

Rhizomes, Barks, Floral Parts, Buds, Fruits and Seeds

I. ALLSPICE

Pimenta officinalis Lindl
Family: *Myrtaceae*

[ALLSPICE, PIMENTO, JAMAICA PIMENTO, JAMAICA PEPPER, CLOVE PEPPER]

GENERAL

An evergreen tree, 25 to 30 feet in height; trunk erect, bark gray; much branched; round-topped; foliage dense. Leaves opposite, oval-oblong to elliptical, entire, glandular-punctate on lower surface; 4 to 6 inches in length; deep green, lustrous. Flowers borne on racemose cymes, small, white to greenish-white, sepals 4, petals 4, stamens numerous; ovary 2-loculed, inferior.

Allspice fruits are harvested when mature but still green. The fruits are dried in the sun; their color changes during drying to dark reddish-brown.

Allspice trees bear fruits in about 7 years and are productive for about a further 12 years.

MORPHOLOGY

A dried, dark reddish-brown, hard berry with a rough surface caused by numerous protuberant oil cavities underlying the exocarp. Calyx and style remains at the apex, stalk scar at the base or, occasionally, a short stalk or portion of stalk. In shape, nearly globular. In size, from about 4 to 7 mm. Pericarp woody, brittle, and around 1 mm in thickness.

The berry has 2 locules separated by a thin dissepiment or partition. Each locule contains a single, reniform, hard, dark brown seed. The embryo is large, curled, with prominent radicle and small cotyledons.

The aroma is fragrant, clove-like; the taste is clove-like, strongly aromatic, pungent.

The name "allspice" is given to this spice because its flavor is said to resemble a combination of cinnamon, nutmeg and cloves.

HISTOLOGY

Pericarp

This consists of exocarp, mesocarp, and endocarp.

Exocarp. In cross section, consisting of epidermis with stomata and short, thick-walled, unicellular, nonglandular hairs up to about 150 microns in length. These hairs are easily broken off in the preparation of the section and are best studied in surface view.

The cells of the epidermis are small, up to about 10 microns, long axis; tangentially oblong-rectangular or nearly so, and have strongly cutinized outer walls. In surface view the epidermal cells are small, from about 5 to 22 microns, and polygonal in shape.

Mesocarp. Cross section. In the outer mesocarp, near the epidermis, there is an oil zone consisting of numerous cavities in which the essential oil of allspice is secreted. These cavities are oval or round in shape. The oval cavities vary in size to about 210 microns, long axis, and to about 150 microns, short axis. The round cavities vary in size from about 60 to 210 microns in diameter; the majority are from about 150 to 180 microns in diameter.

Scattered throughout the mesocarp are numerous pitted, striated stone cells varying greatly in amount of wall thickening, in size and shape. They occur singly, in small groups, and in masses. Individual cells reach up to about 110 microns, long axis, and in the apex region of the mesocarp to about 150 microns; the latter stone cells have, in general, thinner walls.

Fibrovascular bundles occur in the mesocarp. Seen in longitudinal section the vessels are spiral; the fibers have thick, pitted walls, and vary in width and length. The fibers are not numerous.

The ground tissue of the mesocarp consists of thin-walled parenchyma cells, somewhat compressed, and irregular in shape. Rosette crystals of calcium oxalate are scattered throughout the mesocarp but occur more abundantly near and between the oil cavities.

Groups of stone cells occur in the inner mesocarp. These cells vary in size from about 30 to 60 microns, long axis, and the groups extend to about 1300 microns in length. A few layers of compressed parenchyma cells separate the inner mesocarp stone cells from the endocarp.

Endocarp. The endocarp consists for the most part of a layer of compressed parenchyma tissue. In the apex region, near the locule partition in the area of the calyx remains, the endocarp cells become rectangular in shape, and average about 12 microns square. In surface view, for the most part, the endocarp cells are elongated, narrow, and thin-walled.

Locule Partition

The locule partition or dissepiment consists of parenchyma tissue with a few stone cells and numerous crystals of calcium oxalate; both rosette and prismatic crystals occur. Vascular tissue traverses the parenchyma. A very thin cuticle is present on both surfaces.

Seed

Seed Coat. In cross section, the seed coat consists of compressed parenchyma tissue contained between an outer and inner epidermis except on the outer convex and inner flat sides of the seed where the outer and inner epidermis enclose several layers of cells filled with a rich orange-red, gummy or resinous substance. These cells have been called "port wine" cells by early writers and are frequently referred to as such nowadays. This description might apply, but only approximately so, when the resin cells are clumped together and less penetrable to transmitted light. The resin cells are an important identifying characteristic of ground allspice.

The outer walls of the outer epidermis are strongly cutinized, those of the inner epidermis weakly so. In surface view the cells of both the outer and inner epidermis are elongated, narrow, and have faintly beaded walls. They vary in arrangement, with a tendency to parquetry in places.

Vascular tissue with delicate spiral vessels occurs in the seed coat and is usually seen on the inner side of the resin parenchyma adjoining the inner epidermis.

Embryo. The embryo is curled and consists of a large radicle and two small cotyledons. In cross section, the epidermis consists of small, clear cells covered by a very thin cuticle. Underlying the epidermis there is an oil zone with cavities smaller than, but similar to, those of the pericarp. Rudimentary vascular tissue occurs in the central zone of the embryo; delicate spiral vessels are observed in longitudinal section. The remaining tissue consists of small, isodiametric parenchyma cells containing mostly round but some truncated granules of starch, 4 to 12 microns in diameter, and with central hilum.

Pedicel. An untreated cross section of the pedicel viewed under low power shows the epidermis and cortex tissue dark reddish-brown; phloem tissue and pith dark orange; fibers and xylem obscurely yellow, the fibers masked by the surrounding orange color, and the xylem by light transmitted through the lumen of the radial rows of vessels and accompanying tissues.

For a detailed examination, a section should be decolored and the cells expanded by treatment with bleaching solution. A second section should be so treated and then stained with chlor-zinc-iodine.

In cross section, the pedicel consists of (a) epidermis of small cells with strongly cutinized radial and outer walls; (b) cortex of thick-walled, irregular-shaped cells varying in size to about 67 microns, long axis; intercellular spaces; rosette crystals of calcium oxalate occurring in some cells; and occasional oil cavities, round, oval or elliptical, from about 65 to 85 microns, diameter or long axis; (c) a broken ring of thick-walled pericycle fibers; (d) thick-walled phloem tissue with some cells containing prismatic crystals of calcium oxalate; (e) an unbroken ring of xylem tissue with vessels in radial rows, fibers, parenchyma cells, and xylem ray cells; (f) pith of thick-walled parenchyma cells, some containing rosette crystals of calcium oxalate, and thick-walled fibers singly or in groups.

In surface view, the epidermis consists of rectangular to polygonal (mostly the latter), thick-walled, frequently pitted or beaded cells up to about 25 microns, long axis; numerous hairs, straight or curved, pointed, unicellular, varying in length to about 165 microns. No stomata.

Seen in longitudinal section, the pericycle fibers are long, thick-walled, pitted, with lumen narrow to wide. The pith fibers are shorter than those of the pericycle, pitted, and with narrow lumen. Noteworthy in the phloem tissue are chains of prismatic crystals of calcium oxalate; crystals imperfect, but of tetragonal prism-pyramid, rhombic, and cubic types up to 12 microns, long axis. The xylem consists of numerous spiral vessels, fibers, and thick-walled, pitted parenchyma cells.

Ground Allspice

The most important identifying characteristics of ground allspice are (a) the so-called port wine cells of the seed coat, the resin cells; (b) the stone cells, which are numerous; (c) fragments of the pericarp with oil cavities; (d) fragments of the exocarp, and short unicellular hairs; (e) starch; and (f) fragments of the embryo.

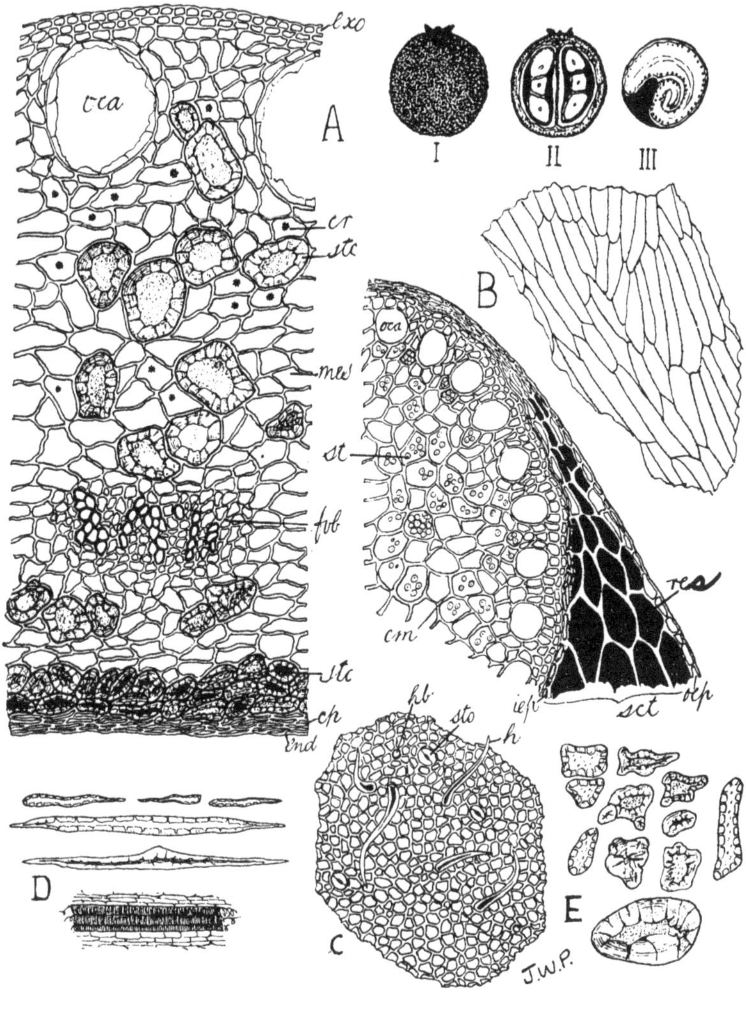

Fig. 1. ALLSPICE. I—Fruit. II—Fruit, longitudinal section. III—Seed, cross section (A) Cross section of pericarp. (B) Cross section of a portion of seed (left): surface view of epidermis of seed coat (right). (C) Exocarp, surface view. (D) Fibers and spiral vessels. E, Stone cells.

cp, compressed cells; cr, crystals; em, embryo; end, endocarp; exo, exocarp; fvb, fibrovascular bundle; h, hair; hb, hair base; iep, inner epidermis; mes, mesocarp; oca, oil cavity; oep, outer epidermis; res, resin cells; sct, seed coat; st, starch; stc, stone cells; sto, stoma.

II. CAPSICUMS

Capsicum
Family: *Solanaceae*

GENERAL

Herbaceous or semiwoody plants, annual or perennial, 1½ to 4 feet or more in height, of many varieties and diverse forms.

Stems branched and more or less angular or channelled.

Leaves alternate, entire, accuminate, petiolate, dark green on the upper surface, paler on the lower.

Flowers axillary, solitary or in fascicles of 2's or 3's; corolla gamopetalous and generally with 5 to 6, sometimes 7, segments, rotate; stamens generally 5 to 6, adnate, and arising from the corolla; ovary superior.

Calyx saucer to cup-shaped, persistant, normally with six short teeth and a thick peduncle.

Fruit a podlike berry, generally with 2 to 4 locules united in the upper part and divided below; varying in shape and size; seeds small, discoid.

Capsicum Spices

Cayenne pepper, red pepper, chilli powder, and paprika are all prepared from dried, ripe fruits of the genus *Capsicum.*

Dried capsicum red peppers vary considerably in size and flavor. The small, conical, acrid peppers are commonly known as chillies; they are used whole for pickling purposes, or they are ground to make Cayenne pepper. Red pepper may be Cayenne pepper, or a pungent pepper made by grinding larger fruits such as the Long Red Cayenne. Chilli powder is prepared from selected, mild or pungent Capsicum fruits, depending upon whether a mild or "hot" product is required.

Cayenne pepper is ascribed to *Capsicum frutescens*, *Capsicum boccatum*, or some other small-fruited species of *capsicum* and paprika to *Capsicum annuum* in government standards, but it should be pointed out that it is the opinion of at least one notable systematic botanist,—namely, L. H. Bailey,—that there is only one species of the genus, the shrubby *C. frutescens*, and that all other horticultural kinds are but forms of this species, including the annual kinds which, he says, develop in a short season and do not become woody before being killed by frost.

MORPHOLOGY

Capsicum fruits employed in the spice trade vary in size from about 6 mm to 15 cm in length. The dried fruits are laterally compressed, conical, base blunt, apex sharp, pointed; calyx and peduncle usually attached to the larger fruits, and sometimes to the smaller; color varying from a murky brownish-red to a rich deep-red.

The pericarp is thin, lustrous; smooth in some varieties, wrinkled in others; containing 2 to 4 locules with central placentae united at the base, and toward the apex parietal placentae marked by longitudinal ridges.

The seeds are numerous, attached to the central placentae or free; discoid, nearly circular, flat with thickened edge; prominent, pointed micropyle adjoining the hilum; hard, smooth, yellow; about 2 to 5 mm in size; and containing a curved embryo embedded in the endosperm.

Odor characteristic, sternutatory, but not unpleasant; taste mildly to intensely pungent.

The more important pungent capsicum fruits are:

African Chillies (Mombasa, Congo, Zanzibar). Length about 9 to 15 mm; murky brownish-red; wrinkled, lustrous, very bright; very acrid. Nigerian chillies: Length 12 to 31 mm; bright red; wrinkled, lustrous, very bright; acrid.

Mexican Chillies (Chiltepin, also known as Chilepiquine, Chili piquin, Chile petine). Length 6 to 14 mm; orange-red; smooth, lustrous, bright; very acrid.

Japanese Chillies (Honka, known as Hontakas). Length 25 to 50 mm; orange-red to dark red; compressed, somewhat wrinkled, lustrous, bright; very pungent.

U. S. A. Domestic Sports (Louisiana sport pepper). Length 25 to 57 mm; orange-red to deep-red; smooth, lustrous, bright; very pungent.

An important large fruit is the **Mexican Chilli Pod** (**Chile Ancho,** var. *accuminatum*). Large fruits, from about 9 to 11 cm in length, and from about 6 to 7 cm at the widest part of the fruit when measured in the dried, compressed state. The pericarp is wrinkled, lustrous, and maroon in color. The calyx is of the flat or saucer type and adheres firmly to the fruit; the peduncle is curved, stout, and from 3.8 to 5 cm long.

Mexican chilli pods are moderately pungent. They are employed in the making of Mexican chilli powder.

An examination of the tissues of Mombasa, Nigerian, Congo (Africa); Piquin (Mexico); Hontaka (Japan); and Louisiana (America) chillies show that these varieties are almost entirely similar in structure. The principal differences lie in the existence of a hypodermis of 1 to 3 layers of cells in Louisiana Sport chillies, an hypodermis of 1 layer of cells in Hontaka chillies, and no hypodermis in Piquin and African chillies; in the thickness and beading of the cell walls of the epidermis as seen in surface view; and, to a limited extent, in the size of some cells.

Paprika is the finely ground product of mature, dried, sound and clean fruits or peppers of *Capsicum annuum* L. A number of types of paprika are manufactured but they can be included in three main categories—namely, Sweet, Semisweet, and Pungent.

There are also various qualities of each type. First quality paprika is prepared from the pericarp of first quality peppers. No peduncles, calices, placenta tissue or seeds are included in the product. Second quality paprika is prepared from the pericarp of good quality peppers and the product may include a small percentage of seeds and adhering placenta tissue but no peduncles or calices. Lower qualities are prepared from peppers of lower quality and include pericarps, placentae, seeds and, in some low grade paprikas, peduncles and calices. For the manufacture of Sweet and Semisweet paprikas, where included, the seeds and adhering placenta tissue are macerated and washed in water to deprive them of part of their pungent content, *capsaicin*.

The color of first quality paprika is a rich red, that of second quality a less intense red. Lower quality paprikas vary in color from brick-red to brownish-red or brownish-yellow.

Paprika is manufactured in a number of countries including the United States of America. In recent times, of the paprika imported by the United States the greatest quantity came from Spain—5,314,497 lb in 1967.

In Spain, three types of paprika are manufactured—namely, "Dulce" or Sweet, "Agridulce" or Semisweet, and "Picante" or Pungent. The last mentioned paprika is prepared from the fruits of *C. annuum* variety *longum*. It is not exported to the United States of America. Each type is marketed in three qualities: Extra, Select, and Common. These paprikas are prepared as above. A limit of 10% is imposed upon seed content in the Select quality and 30% in the Common.

Four varieties of capsicum peppers are employed in the manufacture of the sweet paprikas. As named in Spain, they are Ramillete, Tres Cascos, Bola, and Albar. The fresh fruit of the first variety is globose in shape but somewhat flattish at the base and apex. It has a thick, fleshy pericarp, four locules, and many seeds. The fresh fruit of the second variety is globose but distinctly flat at the apex. It has a less fleshy pericarp, three locules, and numerous seeds. The fresh fruit of the third variety is truncated-conical in shape and is somewhat smaller than the previous varieties. It has a pericarp about half as fleshy as that of the first variety, three locules, and many seeds. The fresh fruit of the fourth variety is globose but flat at the base and apex. The calyx has six lobes as against five in the other varieties. The pericarp is about as fleshy as that of the second variety. It has three large locules, a small, ruddy placenta, and many seeds.

The following table shows the weight of the parts of the fruit at the termination of the drying process, obtained from 100 kilograms of fresh fruit:

		Ramillete	*Tres Cascos*	*Bola*	*Albar*
Pericarps	kg	11.84	10.54	11.30	12.04
Seeds	kg	10.00	9.00	8.54	8.83
Placentae	kg	1.44	1.44	1.52	1.33
Peduncles and calices	kg	1.40	1.54	1.50	1.16
		24.68	22.52	22.86	23.36
Less:					
Peduncles and calices		1.40	1.54	1.50	1.16
Weight of dried pods	kg	23.28	20.98	21.36	22.20
Percentage of pericarps to pods		50.86	50.24	52.90	54.23
seeds " "		42.95	42.90	39.98	39.77
placentae " "		6.19	6.86	7.12	6.00
		100.00	100.00	100.00	100.00

It should not be concluded that a high yield of pericarps, as the Albar variety shows, will necessarily produce a good quality paprika. Color and smoothness are essential properties of paprika. To obtain these and other elements which characterize good paprika advantage must be taken of the particular qualities each variety of peppers can contribute to the product.*

HISTOLOGY

Pericarp

Cross section unless otherwise indicated. The tissues of the pericarp consist of the following:

* D. Ignacio Vivancos Guerao, *Estudio de las Variedades de Pimentos Pimentoneros*, I.N.I.A., Ministerio de Agricultura, Murcia (Spain), n.d.

Cuticle. The cuticle varies in thickness from about 18 to 24 microns in the different varieties of capsicum fruits. In some varieties the cuticle almost surrounds each cell of the epidermis, penetrating down between the cells. In Paprika the cuticle shows distinct grooves, many 12 to 15 microns deep; in surface view these grooves are numerous and of various lengths, many from about 45 to 100 microns long.

Epidermis. The epidermal cells of all varieties tend to a more or less rectangular shape, and are arranged in fairly regular order. There are areas where the cells are of irregular order, and with shapes varying from nearly rectangular to triangular or polygonal, but these cannot be held to be characteristic of any particular variety. The epidermal cells of Congo chillies, measured between the walls, are from 20 to 45 microns, long axis; of Hontaka chillies, from 30 to 45 microns; of Nigerian, Mombasa, Piquin, and Louisiana Sport chillies, from 30 to 60 microns; of the large-fruited Ancho chillies, from 30 to 60 microns; and of Paprika, from 30 to 60 microns. In every variety there are cells that will exceed these measurements but they are exceptional and not typical. The length of the short axis varies somewhat, but in most cases the cells are in the neighborhood of 20 microns.

Seen in surface view, the epidermal cell walls of Piquin, Nigerian and Mombasa chillies are 5 to 6 microns wide; of Congo chillies, 8 microns; of Louisiana Sport chillies, 9 to 12 microns; of Hontaka chillies, 24 to 27 microns; of Ancho chillies, 9 to 12 microns; and of Paprika, 9 to 10 microns. Pitting occurs in the walls of the epidermal cells of these fruits, but some variation exists in the style and extent of pitting in the different varieties. In some varieties the pitting is clearly observable with low power in others only with intermediate, or high, powers. For example, the epidermal cell walls of Congo and Mombasa chillies are similarly pitted, but the pitting of the Mombasa walls becomes apparent at ×250 and higher power, whereas that of the Congo walls is distinct at ×100 power. The pitting does not extend to the middle lamella in Piquin and Hontaka chillies. In Louisiana Sport chillies the pitting frequently extends to the middle lamella and is distinct at ×100 power. Pitting is also a characteristic of the epidermal cell walls of the larger fruits.

The color of the epidermis, observed in surface view, varies from a weak yellow to orange, but this is not a very helpful aid to the recognition of the different varieties of capsicum fruits. The color is affected by exposure to light, and deteriorates to some extent as the chillies or powders age.

Hypodermis. The hypodermis consists of tangentially oval cells with thick walls. In mature Hontaka chillies there is usually a single layer of these cells, occasionally more, varying in size up to about 66 microns, long axis. In Louisiana Sport chillies the hypodermis consists of 1 to 3 layers of cells, up to about

45 microns. In the large-fruited capsicums the hypodermis consists of several layers of cells varying in size from about 30 to 60 microns, long axis, with walls 12 to 15 microns thick.

Mesocarp. The mesocarp consists of parenchyma cells that have been compressed into a tangentially oblong-oval shape. These cells vary greatly in size, extending in the small fruits to 180 microns over the long axis, and in the large fruits to 700 microns. The cell contents consist of chloroplasts and oil; the oil globules are numerous and vary in size.

When brought into contact with concentrated sulfuric acid drawn under the cover glass, the oil drops turn an indigo blue in color.

Vascular Bundles. In all capsicum fruits the vascular bundles of the pericarp consist of xylem tissue with spiral vessels, and phloem tissue. There are no fibers in the pericarp. The spiral vessels are from 9 to 12 microns in diameter for the most part but some reach 21 microns. Vessels of the latter diameter are not abundant. The vascular bundles are situated in the inner region of the mesocarp.

Giant Cells. The so-called "giant cells" are situated between the mesocarp and the thick endocarp, and are bounded on their radial walls by a triangular or wedge-shaped group of parenchyma cells. The giant cells are responsible for the numerous blisters seen on the inner surface of the pericarp. They are characteristic of all capsicum fruits.

Endocarp. The endocarp bounding the giant cells consists of somewhat rectangular or oblong-rectangular cells with thick, lignified, pitted walls. Adjoining the mesocarp tissue between the giant cells the endocarp consists of cells similar in shape to those of the thick endocarp but with thin, cellulose walls. The lignified walls are from about 9 to 12 microns in width; the cellulose walls about 4 or 5 microns. In surface view, both types of endocarp cells are irregular to polygonal in shape, and vary from about 15 to 90 microns, long axis.

Dissepiment

The endocarp continues on to the dissepiment, modifying and passing into epidermis tissue. A thin cuticle covers each epidermis. Fibrovascular tissue occurs in the parenchymatous central zone; spiral vessels are numerous, phloem and xylem fibers are present.

Calyx

In cross section, the calyx consists of the following tissues:

Outer Epidermis. In the small-fruited capsicums the cells of the outer epidermis are mostly rectangular-oblong in shape, fairly regular, with beaded cell walls. They vary in size from about 20 to 75 microns, long axis, and are about 15 microns, short axis. Stomata are numerous. Occasional 1- and 2-celled hairs from about 45 to 90 microns long are present. A thin cuticle covers the epidermis.

In the large-fruited capsicums the cells of the outer epidermis are mostly rectangular-oblong in shape, some are triangular or nearly so, others are polygonal. They vary in size from about 30 to 90 microns, long axis, and are about 24 microns, short axis. The cell walls are beaded and about 6 microns thick. Stomata are numerous, and there are no hairs. A cuticle is present.

Outer Hypodermis. An hypodermis exists in the calyx of both small and large fruits and consists of partially collapsed cells varying in size from about 20 to 30 microns, long axis, in the small fruits, and from about 20 to 200 microns in the large. The cell walls are partially lignified.

Mesophyll. In the calyx of both small and large fruits the mesophyll consists of large, collapsed parenchyma cells with contorted walls. Fibrovascular bundles occur in this tissue.

Inner Hypodermis. Similar to the outer hypodermis, but the cells are smaller.

Inner Epidermis. With the difference that the cell walls tend to be sinuous, the cells of the inner epidermis are similar to those of the outer; in the large fruits these cells are up to about 105 microns, long axis. Stomata are not numerous. In both small and large fruits numerous multicellular, capitate hairs arise from the inner epidermis. These hairs are about 90 microns long; the basal cell is about 30 microns long and about 25 microns in diameter; the head varies from about 30 to 36 microns in diameter. The basal cell is colorless, the remaining cells have yellow to yellowish-orange contents.

Stone Cells

In the region of fusion of calyx and peduncle there are numerous sclerenchymatous cells with thin, pitted walls. The shape of these cells varies considerably; in size they vary from about 65 to 120 microns, long axis.

Peduncle

In cross section the peduncle consists of the following tissues:

Epidermis. In the small-fruited capsicums the cells of the epidermis are mostly rectangular-oblong, or nearly so, in shape; they vary from about 20 to 90 microns, long axis, and from about 15 to 20 microns, short axis. The cell walls are beaded, and are about 4 to 5 microns thick. Stomata are fairly numerous. A thin cuticle is present.

In the large-fruited capsicums the epidermal cells vary in shape; the greater number are rectangular-square or rectagular-oblong, but some tend to the triangular, and some are polygonal. They vary in size from about 20 to 90 microns, long axis, and are from about 20 to 60 microns, short axis; the cell walls are beaded, and from about 4 to 6 microns thick. Stomata are not numerous. A thin cuticle is present.

Hypodermis. In small-fruited capsicums the hypodermis may not exceed a single layer of cells. These cells vary in size from about 12 to 21 microns, long axis, and have thick, regular walls. In the large fruits, the hypodermis may consist of 3 layers of cells, varying in size from about 20 to 75 microns, long axis, and with thick, somewhat contorted walls. In both types of fruit the hypodermal cells are partially lignified and refract polarized light.

Cortex. The parenchyma cells of the cortex appear loose and collapsed; there is no regularity to the shape of the cells, and the walls are much contorted. They vary in size: in the small fruits from about 20 to 60 microns, long axis, and in the large fruits from about 20 to 75 microns with occasional cells reaching as much as 450 microns, long axis.

Fibrovascular Tissue. Phloem fibers occur in close proximity to the outer side of the external phloem tissue. They occur singly, in pairs, and in small groups separated by parenchyma tissue. In the small fruits they vary in diameter from about 9 to 21 microns, with walls from about 3 to 6 microns in thickness; and in the large fruits from about 9 to 84 microns in diameter, with walls from about 4 to 15 microns in thickness.

External Phloem. In both small and large fruits the external phloem of the peduncle consists of a continuous ring of closely arranged cells.

Xylem. In both small and large fruits the xylem forms a continuous ring of tissue broken only by the radial xylem rays. In the small fruits the xylem fibers and the vessels are around 9 or 10 microns in diameter; in the large fruits from about 9 to 30 microns. Spiral, reticulate, and scalariform vessels occur in both small and large fruits; the fibers are pitted.

Internal Phloem. In the small fruits only a few fragments of disintegrated phloem cells remain. In the large fruits the internal phloem consists of a continuous ring of small, closely arranged cells.

Pith. Pith tissue is seen only in the peduncle of large fruits, and then not in its entirety. It is more abundant toward the calyx; away from the calyx the pith is almost entirely absent and the peduncle appears hollow. The pith cells are collapsed, and the cell walls are contorted; they vary in size from about 60 to 160 microns, long axis.

Seed

The seed of all types of capsicum fruits is flat, discoid, and consists of a strong, tough seed coat covered with a thin cuticle; an abundant endosperm, and an embryo with long cotyledons.

Seed Coat. In cross section, this consists of (a) an outer epidermis of large cells with thick, sclerenchymatous basal and radial walls, and a cellulose outer wall thinly lignified on the inner side. The size of the cells and the extent of the thickening is very much greater on the edge or ridge of the seed than on the flat sides (see also under "Ground Capsicum"); (b) a few layers of parenchyma cells, usually collapsed and disintegrated, but sometimes up to 3 layers of sound cells; (c) an inner epidermis of small, regularly arranged, thin-walled cells.

Endosperm. This consists of polygonal cells varying in size from about 10 to 55 microns, long axis, with colorless walls from about 3 to 6 microns in thickness. The cell contents are aleurone grains and globules of fixed oil.

Embryo. This lies coiled and embedded in the endosperm. It is fully developed and consists of radicle and two cotyledons.

Ground Capsicum

Fragments of all the small-fruit tissues described above may appear in the microscope field of cayenne pepper. In some samples, tissues of the calyx and peduncle may be absent; calyx and peduncle remains are common on Hontaka chillies, not uncommon on Mombasa chillies, but of rare occurrence on Nigerian, Congo and Piquin chillies.

In samples of certain high-grade paprikas only large-fruit pericarp tissues will be seen; in other grades these and seed tissues, and in still others, all the tissues of large fruits described above.

An outstanding identifying characteristic of ground capsicum is the sclerenchymatous endocarp; of cayenne pepper and most paprikas, also the thick-walled cells of the outer epidermis of the seed coat. Seen in surface view, the ridge cells of the outer epidermis are greatly contorted, small, and have very thick, sinuous walls. The cells of the flat sides are equally contorted, but very large; the cell walls are sinuous, and have much less thickening than those of the ridge cells; the thickening, in places, is almost entirely on one side of the middle lamella. The structure and sculpturing of the walls of these cells exhibit a striking design.

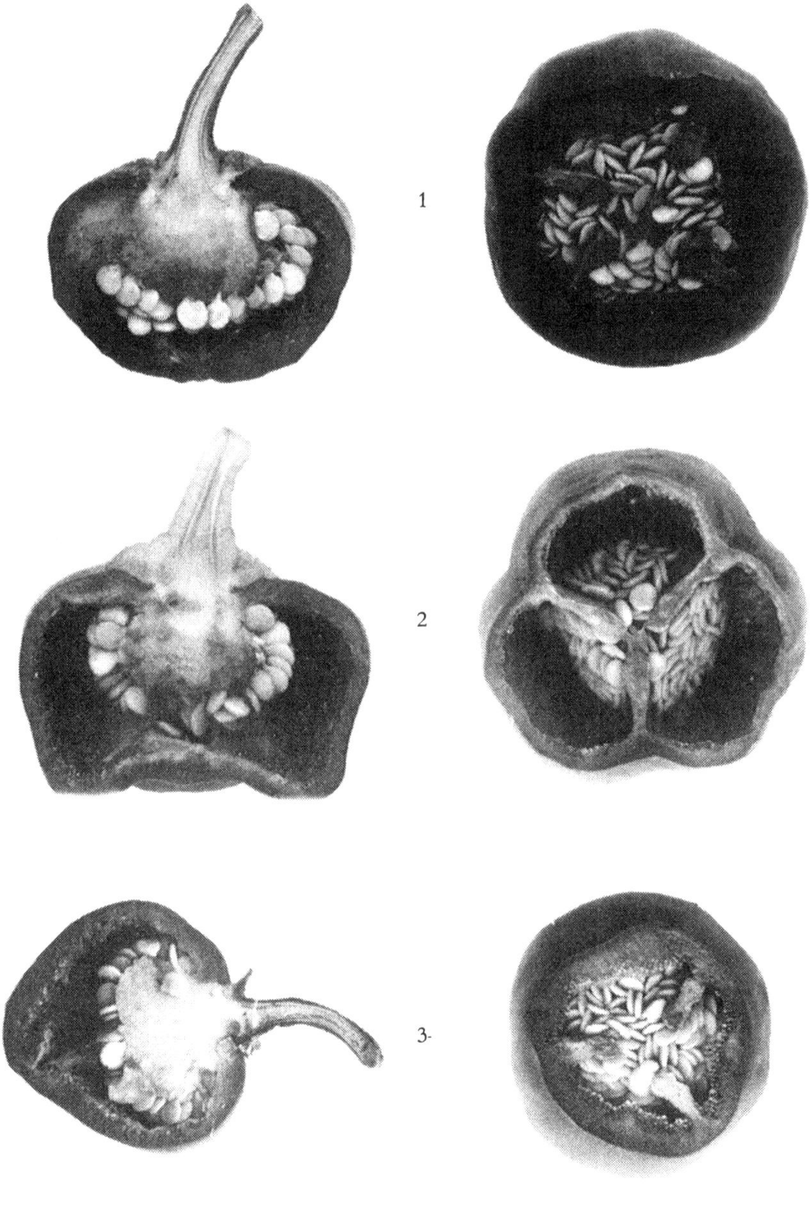

Fig. 2. SPANISH PAPRIKA PODS. 1—Variety *Ramillete.* 2—Variety *Tres Cascos.* 3—Variety *Bola.*

Photographs, Courtesy of the Ministry of Agriculture, Murcia, Spain.

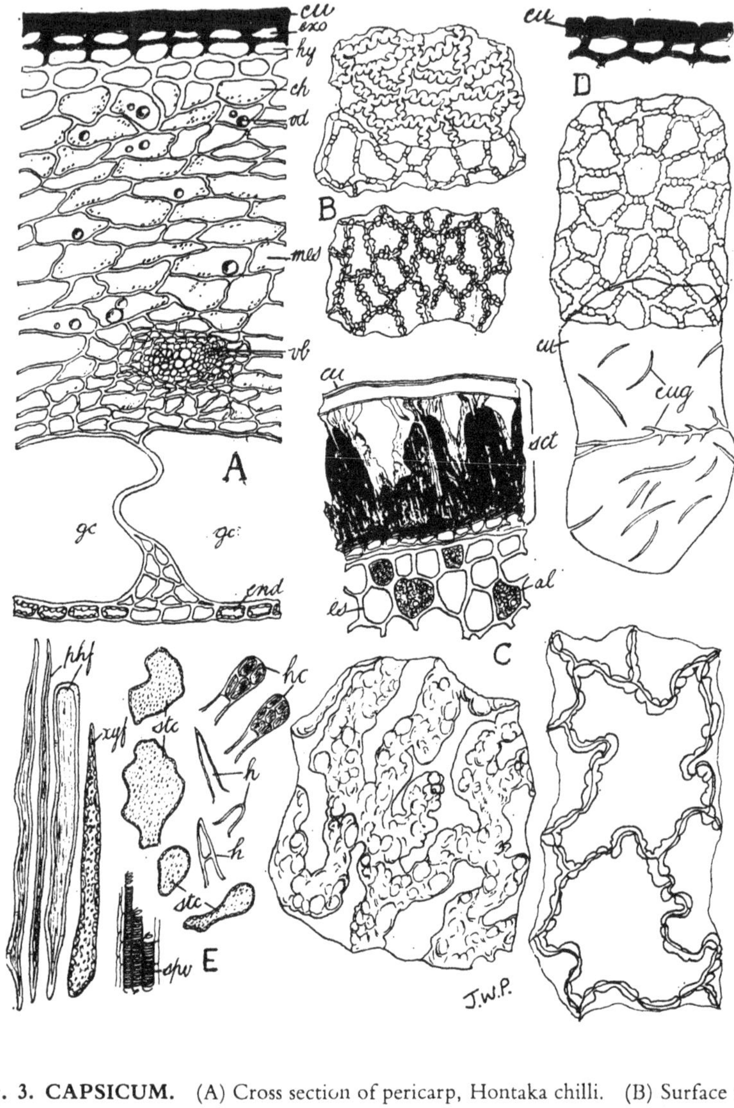

Fig. 3. CAPSICUM. (A) Cross section of pericarp, Hontaka chilli. (B) Surface view of epidermis and hypodermis (above); of endocarp (below). (C) Cross section of portion of seed at ridge (above); surface view of seed coat at ridge (below, left), and of seed coat, flat side of seed (below, right). (D) Paprika, cross section of exocarp illustrating thick cuticle (above); surface view of epidermis and grooved cuticle (below). (E) Fibers, stone cells and hairs of calyx; spiral vessels of fruit.

al, aleurone; ch, chloroplasts; cu, cuticle; cug, cuticle grooves; end, endocarp; es, endosperm; exo, exocarp; gc, giant cell; h, hair; hc, capitate hair; hy, hypodermis; mes, mesocarp; od, oil drops; phf, phloem fibers; sct, seed coat; spv, spiral vessel; stc, stone cells; vb, vascular bundle; xyf, xylem fibers.

III. CARDAMOM

Elettaria cardamomum Maton
Family: *Zingiberaceae*

GENERAL

A perennial, bushy herb reaching a height of about 6 to 12 feet, with a tuberous, horizontal rhizome. Leaves alternate, lanceolate, sheathing, and very long. Flowers arranged in short racemes, corolla tubular, 3-lobed, pale green, androecium with petaloid labellum white in color with pink or purplish veins. Fruit an ovoid capsule, 3-loculed, greenish-brown, and containing several seeds.

The *Mysore* cardamom has large, coarse, dark green leaves with a hard, smooth under-surface. The flower stalks are borne perpendicularly. The fruit is elongated and rarely angled.

The *Malabar* cardamom has large leaves which are silky on the undersurface. The flower stalks lie on the surface of the ground. The fruit is shorter and more globular.

Mysore and Malabar cardamoms are cultivated in southern India and in Ceylon.

The cardamom indigenous to Ceylon is considered to be a variety of the Malabar cardamom. It grows wild in wet forests, and is found especially in the Ratnapura District. Its leaves are coarser than those of the Malabar cardamom and less silky on the under-surface. It has a stem which is distinctly tinged with pink; a short, sparse inflorescence, and a long fruit.

Fully ripe cardamom fruits are a deep yellow in color, but it is the practice to harvest the fruit just before it turns yellow. The fruits are removed individually with the aid of scissors, dried in the sun, clipped to remove pedicel and calyx remains, and sorted and graded according to size and color.

MORPHOLOGY

Cardamom, Dried Unopened Fruit, "Bleached Medium Bold"

Capsule creamy white, oblong-ovoid, 3-sided, from about 8 to 15 mm long, clean, longitudinally wrinkled. Three locules, partitions thin, each locule containing many seeds attached to axile placenta. Pericarp fibrous, papery, and tough.

The unopened capsules have an aromatic odor. The pericarp has a very slight aromatic taste.

Seed

Light reddish-brown to dark reddish-brown, about 3 mm long and from about 2 to 3 mm broad, irregularly 3-sided, angled, transversely wrinkled or furrowed, marked raphe depression, covered by a membranous aril. Perisperm bulky, starchy, white. Endosperm small, pale yellowish-white, of horny consistency and containing a small embryo.

Aroma pleasant and aromatic; taste characteristic, warm, slightly pungent and highly aromatic.

HISTOLOGY

Pericarp

Cross section.

Exocarp. Consisting of epidermis with cells varying in size to about 30 microns, long axis. The epidermis is interrupted in places, probably as a result of damage in the bleaching process.

Mesocarp. Consisting of (a) thin-walled parenchyma cells, irregular but tending to isodiametric or elliptical in shape, and varying in size to about 180 microns, long axis; (b) fibrovascular bundles; (c) occasional bundles of fibers unassociated with vascular tissue; and (d) occasional resin cells.

Endocarp. Consisting of somewhat compressed elliptical or elongated cells. Where the tissues merge into thin locule partitions the cells are small and irregular in shape. The walls of the endocarp cells are slightly thicker than those of the mesocarp cells.

In surface view, the endocarp cells are polygonal in shape with unevenly thickened walls.

Fibrovascular Bundles. The lumen of the fibers varies from about 4 to 66 microns; the walls are thick and pitted. The vessels are spiral, and vary in size to about 50 microns, diameter.

Cell Inclusions. A conspicuous inclusion of the mesocarp parenchyma cells is calcium oxalate. In each cell there occurs 1 or more cystals, solitary, overlying each other when more than one, or in groups, or scattered. Prismatic crystals and crystal sand of calcium oxalate occur in the cells of the locule partitions. Each resin cell contains a lump of resin, pale to dark brown in color, and varying in size from about 44 to 80 microns, long axis.

Seed

Cross section.

Aril. Consisting of 3 layers of nearly rectangular, thin-walled cells containing a pale-colored substance and crystals of calcium oxalate.

In surface view, the cells are greatly elongated.

Seed Coat. Consisting of (a) an epidermis with rectangular cells varying in size to about 25 microns, long axis; (b) collapsed parenchymatous tissue; (c) a layer of large rectangular cells containing globules of oil; (d) compressed parenchymatous tissue; and (e) a conspicuous layer of sclerenchymatous, thick-walled, dark brown, beaker cells, each with a cap or nodule of silica. These cells vary in size to about 33 microns, radial axis, and have an average width of about 16 microns.

In surface view, the cells of the epidermis are elongated, narrow, and have moderately thick walls; the collapsed parenchyma consists of cross cells varying in size to about 135 microns, long axis; the cells of the compressed parenchyma tissue vary in size to about 65 microns, long axis, and have wavy, beaded walls; and the beaker cells are polygonal in shape with the embedded silica clearly visible.

Perisperm. Consisting of somewhat wavy-walled, polygonal cells varying in size to about 88 microns, long axis, and filled with starch. The starch granules are polygonal and vary in size from about 2 to 4 microns; most granules are about 3 microns. Embedded in the starch mass of each cell is a group of calcium oxalate crystals.

Endosperm. The endosperm cells are polygonal with thinner and straighter walls than those of the perisperm cells, and contain a protein substance having a waxy or horny nature. The cells vary in size to about 62 microns, long axis.

Embryo. The embryo is straight. The cells contain aleurone grains and fixed oil.

Ground Whole Fruits of Cardamom

The most important identifying tissues and cell inclusions are (a) spiral vessels and closely grouped fibers of the fibrovascular bundles of the pericarp; (b) parenchyma cells of the mesocarp with crystals of calcium oxalate; (c) resin lumps; (d) elongated, narrow cells of the aril with pale-colored contents and crystals of calcium oxalate; (e) large parenchyma cells containing globules of oil; (f) dark-brown beaker cells of the seed coat with or without silica nodular caps; (g) starch-filled perisperm cells with embedded crystals of calcium oxalate; (h) endosperm cells filled with protein substance; and (i) scattered crystals of calcium oxalate, particles of silica, starch masses, and aleurone grains.

Ground Cardamom Seed

As above with the exception of a, b, and c.

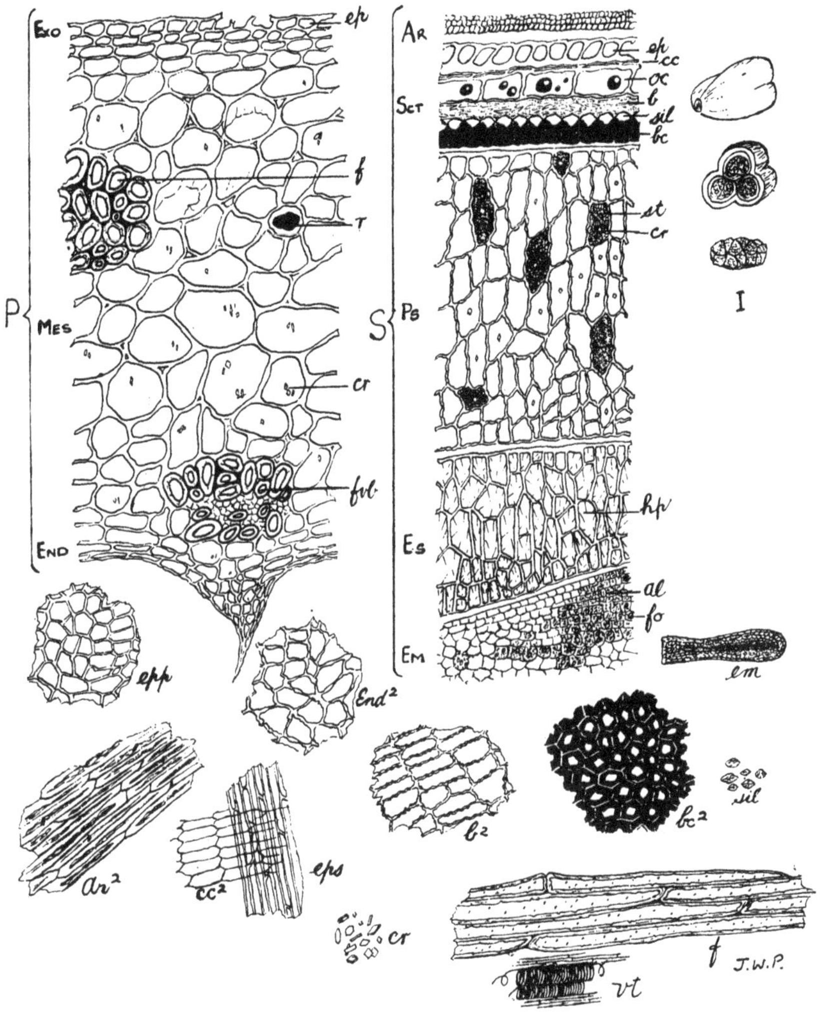

Fig. 4. CARDAMOM. (I) 3-loculed fruit and seeds. (P) Cross section of pericarp. (S) Cross section of seed.

Al, aleurone; Ar, aril (ar², surface view); b, beaded cells (b², surface view); bc, beaker cells (bc², surface view); cc, cross cells (cc², surface view); cr, crystals; em, embryo; End, endocarp (End², surface view); ep, epidermis; epp, epidermis of pericarp, surface view; eps, epidermis of seed coat, surface view; Es, endosperm; Exo, exocarp; f, fibers; fo, fixed oil; fvb, fibrovascular bundle; hp, horny protein; Mes, mesocarp; oc, oil cells; Ps, perisperm; r, resin; Sct, seed coat; sil, silica nodules; st, starch; vt, vascular tissue.

IV. CINNAMON

Cinnamomum zeylanicum Nees
Family: *Lauraceae*

[CEYLON CINNAMON]

GENERAL

The wild cinnamon tree is found in the wet forests of Ceylon up to an elevation of 5,000 feet. It reaches 30 to 50 feet in height; bark thick, rough; branches numerous; leaves reaching about 17 cm in length, ovate or ovate-oblong, margin entire, tip obtuse, petiolate, leathery; flowers inconspicuous, yellow.

Under cultivation, Ceylon cinnamon is kept to bush-like proportions, not higher than 6 to 8 feet. The best cinnamon is grown in the quartz or white sand of the Negombo District, and is generally propagated from seed sown in nurseries. Seedlings are transplanted in the field when about 10 months old, usually when the monsoon rains begin. In 2 or 3 years the shoots are ready for cutting but it is usual to wait until the fourth year to take the first crop. After cutting, more shoots arise providing a succession of wood. The yield diminishes after the tenth year. Cutting takes place twice a year and the greater part of the crop is from 2 year old shoots.

The bark is twice slit longitudinally, one cut on each side of the shoot, permitting its removal in two strips of similar dimension. The pieces are then put together, the outer side of one piece against the inner side of the other, bundled, closely bound together, and allowed to "ferment" for 24 hours. Following this process each length of bark is placed over a cylindrical stick and the outer bark removed with the aid of a curved knife. After a short interval the pieces of bark are placed one within the other and left to dry. As the bark dries it contracts, curls inwards, and gradually acquires the form of a "pipe" or "quill." The bark is graded according to texture and color before quilling. First quality Ceylon cinnamon quills are obtained from the thin bark of tender shoots.

MORPHOLOGY

Ceylon Cinnamon

Compound Quills, Grade "0000." Yellowish-brown, cylindrical, compound quills, approximately 1 m in length and 10 mm in diameter. Scraped, smooth, thin, rolled bark with occasional scars. Fine, light-colored wavy lines running lengthwise and clearly visible. A pleasing, fragrant odor and a warm, sweet, aromatic taste.

Ceylon Cinnamon Quillings. Quillings are broken pieces of compound quills of various grades. The pieces vary considerably in size, being about 5 to 15 or 20 cm in length and about 10 to 25 mm in diameter. Many of the quillings are completely broken down. For aroma and taste they approach but lack the fullness of that of the sound Ceylon cinnamon quills.

Ceylon Cinnamon Featherings. Featherings are short shavings and small pieces of bark left over in the processing of the inner bark into quills. Collectively, featherings present a shade darker color than the quills and a shade lighter than the chips. Featherings have a flavor similar to that of the quillings.

Ceylon Cinnamon Chips. Mature, unpeelable bark is utilized in making "chips." Cinnamon chips, as the name implies, are small pieces of bark; they are grayish-brown on the outside and a lighter brown on the inside. Ceylon chips are deficient in both aroma and taste and are not to be compared to the quills for flavor.

HISTOLOGY

In the preparation of Ceylon cinnamon bark the cork and cortex tissues are removed leaving nothing to the outside of the pericycle except a few fragments of cortex cell walls.

In cross section, the bark consists of (a) pericycle; and (b) phloem. The pericycle tissue consists of stone cells and fibers. The phloem tissue consists of (a) parenchyma cells containing starch; (b) sieve tubes; (c) fibers; (d) oil secretion cells; and (e) phloem ray cells containing prismatic crystals of calcium oxalate.

Pericycle

The stone cells of the pericycle are abundant and form an almost unbroken ring; they vary in shape, and in size to about 150 microns, long axis. The fibers occur in groups among the stone cells.

Phloem

The phloem parenchyma cells are irregular in shape, thin-walled, and contain starch. The starch granules are mostly round in shape but some are truncated; they vary from about 6 to 9 microns in diameter with the former size predominating; the hilum is central.

The sieve tubes are best examined in longitudinal section under medium power. This permits a satisfactory study of the cells and the sieve plates.

Phloem fibers are numerous. Seen in longitudinal section the phloem fibers vary in length from about 100 to 700 microns, and in width from about 15 to 31 microns, with the majority being from about 350 to 550 microns in length, and from about 22 to 28 microns in width at the widest part of the fiber.

The oil cells, in cross section, are oval or nearly round, and vary in size from about 40 to 80 microns, long axis or diameter. In longitudinal section the oil cells are oval and vary in size from about 90 to 150 microns, long axis.

The phloem rays consist of radial rows of somewhat rectangular parenchymatous cells containg small, prismatic crystals of calcium oxalate.

Ground Ceylon Cinnamon

The most important identifying characteristics of ground Ceylon cinnamon are (a) numerous phloem fibers; (b) numerous stone cells; and (c) starch, smaller and less abundant than that of cassia; granules occurring singly and in small aggregates; clearly visible with transmitted light, poorly refractive to polarized light.

With polarized light the fibers and stone cells exhibit striking color and brilliance, appearing white, yellow, reddish-brown, or turquoise blue. Rotating the analyzer will bring these colors into view.

V. CASSIA

Family:—*Lauraceae*

MORPHOLOGY

China Cassia (*Cinnamomum cassia* Blume)

Rolled bark in pieces of varying lengths from about 25 to 38 cm. Most China cassia bark is about 3 mm thick but it varies from about 2 to 6 mm. Unscraped China cassia is gray or grayish; surface rough and irregular, coarse in appearance. Scraped China cassia is light reddish-brown with occasional patches of gray; smooth or nearly smooth.

The aroma of China cassia is faintly aromatic; the taste is sweet, aromatic, pungent, and slightly astringent.

Saigon Cassia (*Cinnamomum loureirii* Nees)

Grayish-brown bark in single or double quills varying in length from about 15 to 30 cm; in diameter from about 10 to 38 mm, and in thickness from about 1 to 6 mm. The thin bark is slightly rough and shows longitudinal wavy ridges, scars marking former branch joints, and numerous wart-like protuberances. The thick bark is grayer than the thin, very rough, and with no wavy ridges.

The aroma of Saigon cassia is agreeable and aromatic; the taste is sweet, aromatic, pungent, and slightly astringent.

Batavia (Indonesia) Cassia (*Cinamomum burmanni* Blume)

Cylindrical double quills of reddish-brown bark varying in thickness from about 1 to 5 mm. The quills are about 13 mm in diameter and of various lengths depending upon the grade, long or short stick. The bark is comparatively smooth. Scars marking former branch joints are apparent, but no wavy lines or ridges. The quills of high quality Batavia cassia are of good appearance, straight and regular; those of inferior grades are frequently twisted and irregular.

The odor of Batavia cassia is agreeable and aromatic; the taste is sweet and pungent.

Batavia cassia is frequently referred to as Batavia cinnamon.

HISTOLOGY

China Cassia

In cross section, the unscraped bark of China cassia consists of (a) cuticle; (b) epidermis; (c) cork; (d) cortex; (e) pericycle; and (f) phloem. (In scraped cassia the cuticle, cork, and more or less of the cortex tissue is absent.)

Cuticle. The cuticle is well-developed and varies in thickness.

Epidermis. The cells of the epidermis are tangentially oblong-rectangular; have dark contents, and vary in size from about 10 to 25 microns, long axis.

Cork. The cork tissue consists of stone cells, thin-walled cells, and arch cells. The thickening of the stone cells is laid down fairly evenly; that of the arch cells occurs on the outer wall, causing the archlike appearance. The thin-walled cells of the cork layer are disorganized.

Cortex. The parenchyma cells of the cortex are irregular in shape, and vary in size; the cell walls are about 4 or 5 microns thick. Intercellular spaces are frequent. Stone cells are numerous in the cortex, and occur singly and in groups. They vary in shape, and in size from about 25 to 150 microns, long axis. Many are characterized by the thinness of one wall, frequently the outer wall. Starch granules occur in the parenchyma cells and in some stone cells.

Pericycle. Pericycle stone cells are numerous, and occur in groups; they do not form an unbroken ring. The walls of these stone cells are, for the most part, more evenly thickened than those of the stone cells occurring in the cortex. The fibers are similar to the phloem fibers but somewhat thicker. They occur in groups but are not abundant.

Phloem. Phloem fibers are not numerous. In longitudinal section, the phloem fibers vary in size but are mostly from about 440 to 690 microns in length, and from about 27 to 35 microns in width. The sieve tubes, in cross section, are irregularly compressed, but are easily indentified in longitudinal section. The parenchyma cells are irregular in shape; the walls are about 4 to 5 microns thick; starch is present in these cells. The phloem rays consist of radial rows of rectangular, parenchymatous cells containing small, prismatic crystals of calcium oxalate.

The *starch* of China cassia consists of round granules, some truncated; cen-

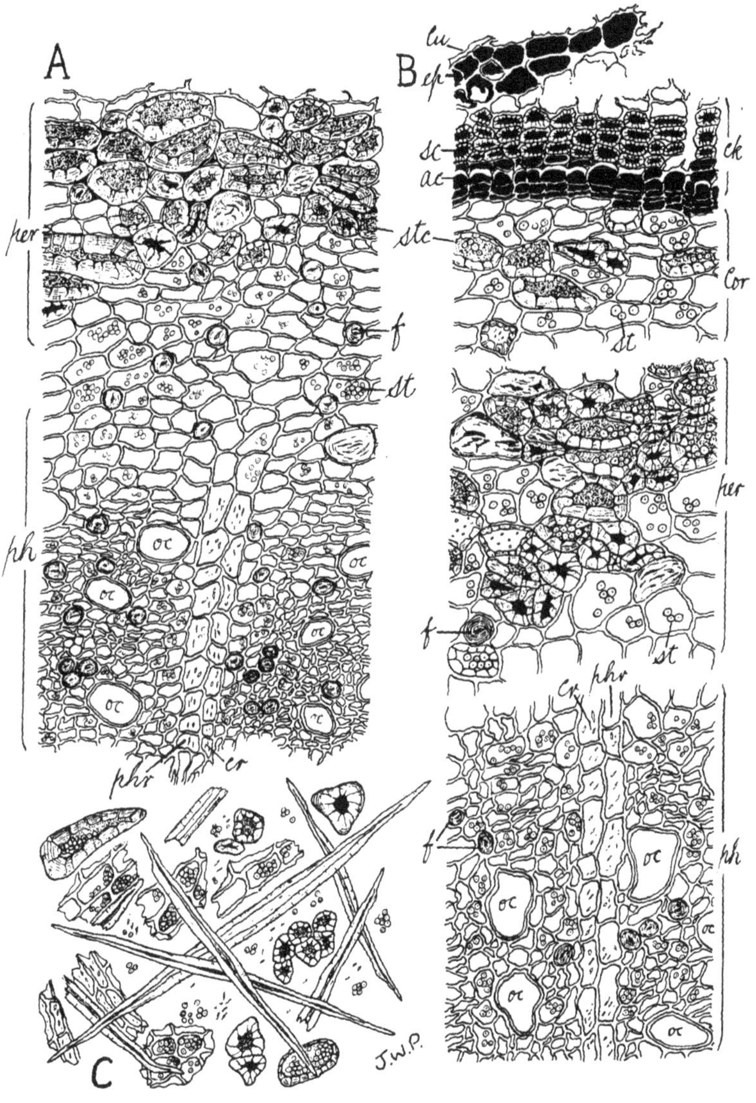

Fig. 5. CINNAMON AND CASSIA. (A) Ceylon cinnamon bark, cross section. (B) China cassia (Kwangsi) bark, cross section. (C) Ground Ceylon cinnamon, illustrating fibers, stone cells, fragments of parenchyma and ray cells, starch, and crystals.

ac, arch cells; ck, cork; cor, cortex; cr, crystals; cu, cuticle; ep, epidermis; f, fibers; oc, oil cells; per, pericycle; ph, phloem; phr, phloem ray; sc, stone cork; st, starch, stc, stone cells.

tral hilum; and varying in size to about 15 microns in diameter. With polarized light the granules exhibit a clear, distinct cross.

Saigon Cassia

The structure of Saigon cassia is similar to that of China cassia with the following differences:

(1) Pericycle stone cells are generally larger with fairly even thickening. Elongated, thick-walled stone cells up to about 225 microns, long axis, occur in groups and are characteristic of Saigon cassia.
(2) The oil cells are larger.
(3) Phloem fibers are fewer and reach greater widths.
(4) The cell contents are dense and dark, particularly in the thick bark.
(5) Starch granules vary in size to about 22 microns diameter in the thick bark; to about 15 microns in the medium; and to about 12 microns in the thin.
(6) The crystals in the ray cells are larger in the thick bark; somewhat larger in the medium; similar in size in the thin but mostly needlelike.

Batavia (Indonesia) Cassia

The structure of Batavia cassia is similar to that of China cassia with the following differences:

(1) The starch granules vary in size to about 22 microns diameter.
(2) Ray crystals are mostly oblong, varying in size to about 14 microns, long axis, and 6 microns, short axis.

Ground Cassia

Samples of China, Saigon, and Batavia cassia will show under the microscope fragments of the tissues just described. A sample of ground cassia might show the fragmented tissues of all three, for it is the general practice to blend these barks in the preparation of ground cassia. Assuming a blended product, the most important identifying characteristics of ground cassia are (a) numerous stone cells, some very large, and many containing starch; (b) fibers, not numerous but up to about 45 microns at their widest part and occasionally larger; (c) abundant starch, with granules up to about 22 microns diameter; round, some truncated; hilum central; occurring singly and in small aggregates; clearly visible with ordinary transmitted light; exhibiting a bright cross with polarized light; and (d) numerous prismatic crystals of calcium oxalate.

VI.—CASSIA BUDS

MORPHOLOGY

Cassia Buds are the dried, unripe fruits of *Cinnamomum cassia* Blume, and *Cinnamomum loureirii* Nees. They vary from about 6 to 14 mm in length, the width of the cup varying from about 4 to 6 mm. The immature fruit is snugly enclosed within a grayish-brown calyx with only the brown, dome-like upper surface of the pericarp visible. The exposed pericarp is smooth and lustrous, with the small, protuberant base of the style in the center. The hidden pericarp, seen when removed from the calyx, is markedly wrinkled. The calyx is cup-shaped, expanding from the calyx tube; rough, wrinkled, very hard, and frequently with pedicel attached.

Cassia buds have a slight cinnamon-like odor, and a sweet, warm, pungent taste akin to that of cassia bark.

HISTOLOGY

[LONGITUDINAL SECTION THROUGH CALYX, PERICARP, AND SEED COAT]

Calyx

Calyx Lobes. The lobes of the calyx consist of the following tissues:

(a) An outer epidermis of small cells, rectangular or nearly so, with thick outer walls. In surface view, these cells are polygonal, thick-walled, and divided by thin cross walls. Hairs are not present.

(b) A cortex of parenchyma cells, secretion cells, and occasional stone cells. The parenchyma cells vary in size to about 75 microns; the secretion cells to about 66 microns, and the stone cells to about 80 microns, long axis.

(c) Fibrovascular tissue consisting of short and long fibers, phloem, xylem, and phloem and xylem parenchyma. The short fibers are porous with numerous pits; they vary considerably in shape and size. The long fibers are pitted and striated; they vary in length to about 1500 microns, and in width to about 36 microns. The phloem cells are numerous. The xylem vessels are small, and consist of annular, spiral, and reticulate types.

(d) An inner epidermis of small, rectangular cells; in surface view similar to that of the outer epidermis.

Calyx Tube. The calyx tube consists of the following tissues:

(a) An epidermis of small cells, rectangular or nearly so, with thick outer walls. In surface view, the epidermal cells are polygonal with thick parallel walls and thin cross walls. Hairs are present. The hairs are unicellular, straight or curved, some with curled tip, and up to about 156 microns in length.

(b) A cortex of parenchyma cells, isodiametric, to about 45 microns diameter; secretion cells to about 66 microns, and occasional stone cells. The cell contents are light to dark brown in color. Starch is present but not abundant; the granules are small, to about 6 microns. The presence of starch is masked by the dark cell contents but it is readily seen after bleaching the section and staining with iodine. Small prismatic crystals of calcium oxalate are distributed throughout the cortex.

(c) A pericycle marked by a ring of fibers. The ring is broken by the parenchymatous tissue of the pith rays. The fibers are thick-walled and up to about 66 microns in diameter.

(d) Fibrovascular tissue similar to that of the lobes described above.

(e) Pith of isodiametric, parenchymatous cells and occasional stone cells.

Pericarp

The pericarp consists of the following tissues:

(a) A sclerenchymatous epicarp. The cells have thick outer walls and thin inner walls. The sclerenchymatous epicarp forms a continuous tissue except where it is interrupted by the ovary canal at the apex of the fruit, and where it merges with the parenchymatous tissue of the calyx at the base of the fruit. The radial thickness of the sclerenchymatous layer varies from about 45 to 58 microns.

(b) A mesocarp of stone cells, parenchyma cells, secretion cells, and vascular tissue. The stone cells vary in size and shape; they underlie the epicarp and are most numerous in the regions adjoining the ovary canal. The parenchyma cells occupy the major part of the mesocarp; they tend to be elliptical or isodiametric in shape and they vary in size to about 66 microns. The cell contents vary in color from light to dark brown. Small prismatic crystals of calcium oxalate are distributed throughout the parenchymatous tissue. The secretion cells are numerous.

(c) An endocarp of sclerenchymatous tissue except in the basal region of the fruit where it is parenchymatous. The endocarp cells in the region of the ovary canal are large and thick, almost without lumen. The remaining sclerenchymatous endocarp consists of more regular, rectangular-shaped cells which contain large prismatic crystals of calcium oxalate.

Seed Coat

The seed coat consists of the following tissues:

(a) An outer tissue of deeply pigmented (dark-brown) parenchyma cells. There are about 6 layers of cells in the area facing the ovary canal of the pericarp, and about 4 layers on the sides. The cells are tangentially elongated to about 90 microns. The outer walls of the cells of the outermost layer are somewhat thickened. In the basal region the cells are less heavily pigmented and become more or less isodiametric in shape.

(b) An almost complete tissue of reticulated cells. In the upper part of the seed these cells are arranged in palisade fashion, the innermost cells being the longest, up to about 100 m microns radial axis. On the sides the reticulated cells are all much the same and up to about 33 microns, long axis. Reticulated cells are not evident in the basal region.

(c) An incomplete, parenchymatous tissue underlying the reticulate cells. Five or 6 layers of irregular-shaped cells are evident. In size, these cells become larger in three successive stages. The outer layer consists of cells to about 28 microns, the next or second layer consists of cells to about 36 microns, and the cells of the remaining layers are up to about 75 microns, long axis. The walls of the outer and second layers of cells are thinner than those of the remaining cells of this tissue. In surface view, the cells of this tissue are polygonal in shape. Calcium oxalate crystals are present, and with polarized light appear as 4-pointed, or more, rosettes varying in size to about 10 microns. Other seed tissues are not present.

Pedicel

In cross section the pedicel consists of the following tissues:

(a) Epidermis. The epidermal cells are small and rectangular with thick outer walls. Hairs are present. In surface view, the epidermal cells are quadrilateral to oblong-rectangular or nearly so, thick-walled, and divided by cross walls either thick or thin. The cells vary in size to about 30 microns, long axis.

(b) Cortex. The cortex consists of isodiametric, thin-walled parenchyma cells with dark contents; secretion cells, and occasional stone cells.

(c) Pericycle. Marked by an almost continuous ring of fibers, small but thick-walled.

(d) An almost complete ring of fibrovascular tissue.

(e) A pith of isodiametric parenchyma cells. Pith rays radiate outwards and traverse the fibrovascular tissue and the pericycle fiber zone.

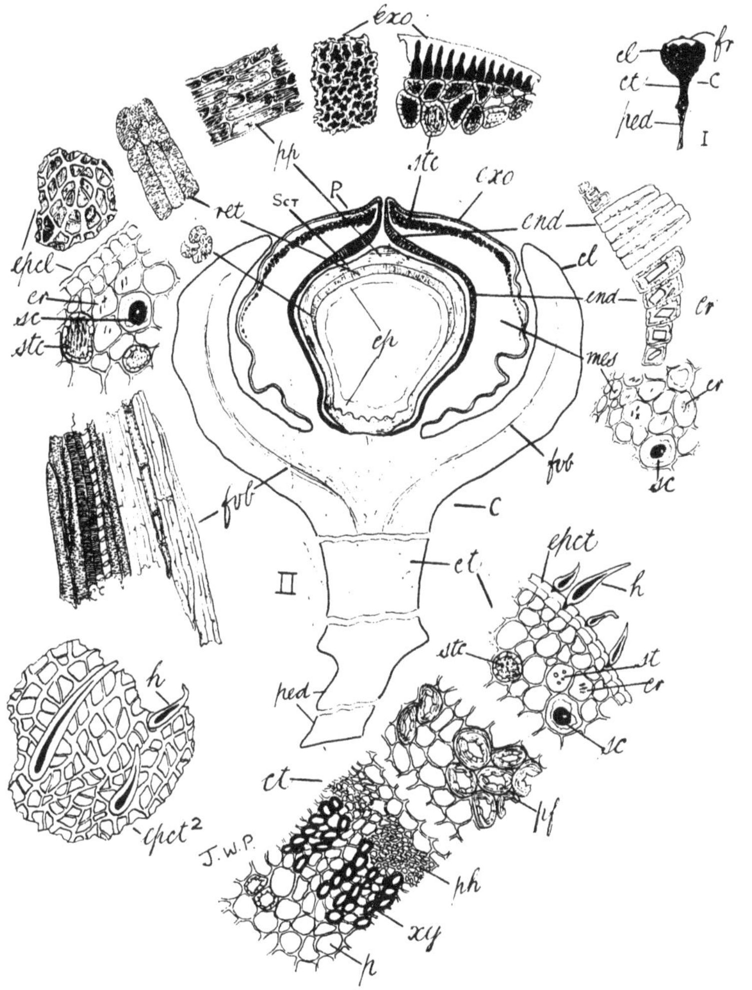

Fig. 6. CASSIA BUDS. I—Cassia bud. II—Longitudinal section of calyx and fruit.

c, calyx; cl, calyx lobe; cp, compressed parenchyma; cr, crystals; ct, calyx tube; end, endocarp; epcl, epidermis of calyx lobe; epct, epidermis of calyx tube; exo, exocarp; fr, fruit; fvb, fibrovascular bundle; h, hair; mes, mesocarp; p, pith; P, pericarp; ped, pedicel; pf, pericycle fibers; ph, phloem; pp, pigmented parenchyma; ret, reticulated cells; sc, secretion cells; sct, seed coat; st, starch; stc, stone cells; xy, xylem.

VII. CLOVE

Caryophyllus aromaticus L.
Family: *Myrtaceae*

GENERAL

The clove tree reaches 16 to 40 feet in height at maturity. It is an evergreen, pyramidal, with a thin, smooth, grayish bark. The leaves are opposite, elliptical to obovate-oblong, accuminate at both ends, blade from about 7.5 to 12.5 cm long and 2.5 to 5 cm wide, entire, glandular-punctate, petiolate, smooth, lustrous, and fragrant. Mature leaves are dark green.

The flowers are arranged in paniculate cymes, fragrant, and produced in either of two seasons, July to October, and November to January; petals 4; stamens numerous; ovary inferior; fruit a purple drupe about 25 mm in length and 12 mm in breadth, and usually containing a single seed. The fruit is called "Mother of Cloves."

The cloves of the spice trade are the unopened flower buds, picked when they turn red at the base. The flower stalks are the "clove stems" of commerce.

Khoker cloves are those buds which under bad drying and storage conditions become darker, rather musty, and finally pale and wrinkled. Cloves are graded according to appearance and impurities present. Zanzibar cloves are graded Special, No. 1, No. 2, and No. 3, as follows:

	Extraneous Matter (stems, mother of cloves, foreign & inferior matter)	*Khoker*	*Moisture*
	%	%	%
Special Grade	3	2	16
Grade No. 1	5	3	16
Grade No. 2	5	7	16
Grade No. 3	5	20	16

MORPHOLOGY

In shape, the clove is said to resemble a round headed nail; the word clove is derived from the French "clou," Latin "clavus," a nail. Cloves are reddish-brown in color, vary from about 12 to 19 mm in length, and consist of hypanthium, receptacle, calyx of 4 calyx teeth, corolla of 4 unopened petals enclosing an indefinite number of stamens and a central columnar style. The ovary, situated just below the receptacle, is inferior, 2-loculed, with central placenta and numerous ovules. The hypanthium proper is that part of the fruit axis below the calyx and ovary, an extension or enlargment of the recptacle without loculi. The entire portion of the fruit axis below the calyx is sometimes called the gynophore, calyx tube, stemlike receptacle, or receptacle. In some cloves the hypanthium is clearly or obscurely 4-sided, in others it is compressed. Sound dried cloves are bold, plump, somewhat rough to the touch but not wrinkled or shrivelled.

Dried cloves have a very strong aromatic odor, and a hot, pungent, aromatic taste.

HISTOLOGY

Hypanthium (a)

Cross section through the ovary (considered part of the hypanthium for this purpose). The tissues consist of:

(a) Epidermis of nearly rectangular cells varying in size from about 9 to 15 microns, long axis, with cuticle.

(b) Volatile oil zone consisting of oval or elliptical cavities varying in size from about 112 to 260 microns, long axis, and from about 70 to 190 microns, short axis. The cavities are interspersed with parenchyma cells pressured into a flattened or elongated shape.

(c) Parenchyma cells with moderately thick walls, and varying in size to about 45 microns, long axis.

(d) Zone of fibrovascular tissue; in longitudinal section the bundles consist of spiral vessels; phloem tissue, and thick-walled phloem fibers accompanied by crystal fibers containing rosette crystals of calcium oxalate.

(e) Bilocular ovary with central placenta and numerous ovules.

Hypanthium (b)

Cross section through hypanthium below the ovary.

The tissues are as above from the epidermis to the fibrovasular bundle zone inclusive. The remaining tissues are:

(a) Inward from the fibrovascular bundle zone there is a mass of spongy, thick-walled parenchyma tissue with chains of thick-walled parenchyma cells and large intercellular spaces.

(b) An inner zone of fibrovascular tissue with small vessels and a few phloem fibers.

(c) Pith, consisting of parenchymatous cells, many containing rosette crystals of calcium oxalate.

In surface view, the epidermal cells of the hypanthium are polygonal or irregular in shape. Stomata are numerous.

Corolla

The corolla consists of:

(a) An outer and inner epidermis with small cells.

(b) Mesophyll of parenchyma tissue with rosette crystals of calcium oxalate.

(c) Numerous volatile oil cavities, in size about 60 to 70 microns, long axis.

(d) Vascular tissue.

In surface view, the epidermal cells of the corolla are small, polygonal or irregular in shape, with walls of uneven thickness and occasional beading. The oil cavities are strikingly conspicuous.

Stamens

Consisting of:

(a) Filament, with small, longitudinally oblong-rectangular cells varying in size to about 10 microns, long axis; central vascular bundle; volatile oil cavities; and rosette crystals of calcium oxalate.

(b) Anther. The anthers are noteworthy for their reticulated cells. Rosette crystals of calcium oxalate also occur in the anthers.

(c) Triangular pollen grains, in size from 15 to 21 microns.

In surface view, the reticulated cells of the anthers are noteworthy, as are the masses of rosette crystals of calcium oxalate in the region of dehiscence.

Clove Stems

In cross section consisting of:

(a) Epidermis with rectangular cells varying in size to about 24 microns, long axis, occasionally larger; and cuticle.

(b) Hypodermis, with stone cells thickened on the inner side. This tissue stains yellow with chlor-zinc-iodine after bleaching.

(c) Zone of numerous volatile oil cavities varying in size from about 120 to 160 microns, long axis.

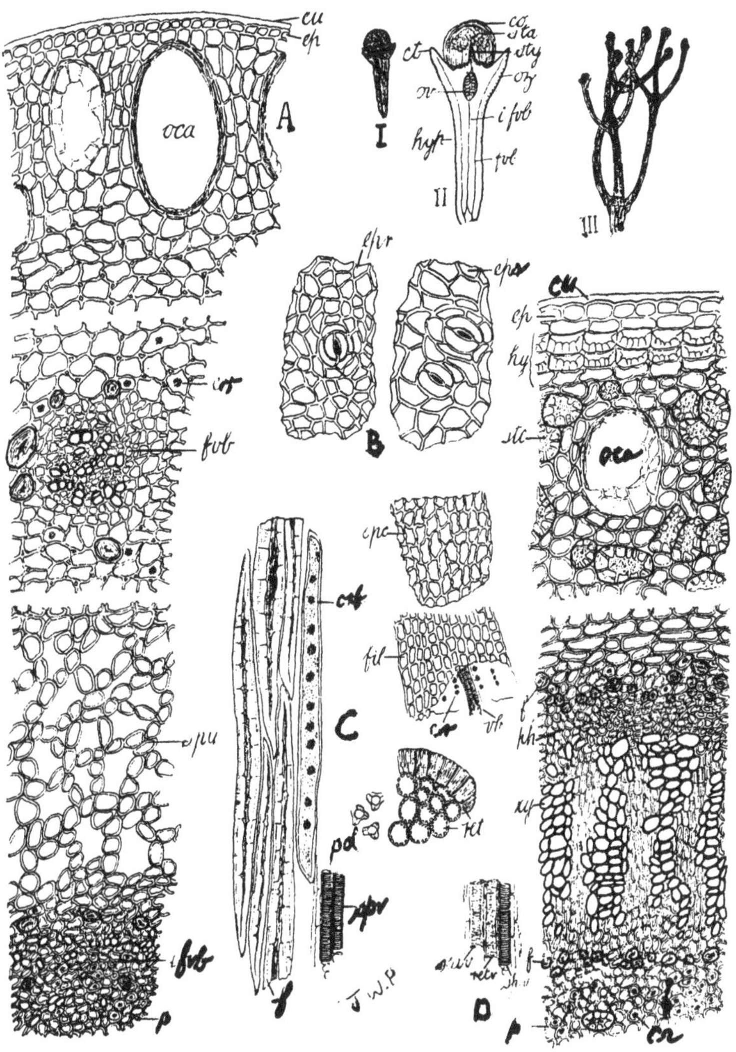

Fig. 7. CLOVE. I—Clove. II—Clove, longitudinal section. III—Clove stems. (A) Cross section of hypanthium below ovary.—(B) Surface view of epidermis of clove (left); of epidermis of stem (right).—(C) Fibers and spiral vessels of clove (left); epidermis of corolla, surface view (above, right); filament (center); reticulated cells of anther (bottom, right).—(D) Cross section of stem (right); vascular tissue (above).

co, corolla; cr, crystal; ct, calyx teeth; cu, cuticle; ep, epidermis; epc, epidermis of corolla; epr, epidermis of hypanthium; eps, epidermis of stem; f, fibers; fil, filament; fvb, fibrovascular bundle; hy, hypodermis; hyp, hypanthium; ifvb, inner fibrovascular bundle; oca, oil cavity; ov, ovary; oz, oil zone; p, pith; ph, phloem; pol, pollen; ret, reticulated cells; retv, recticulated vessel; crf, crystal fiber; scav, scalariform vessel; spv, spiral vessel; spy, spongy parenchyma; sta, stamen; stc, stone cells; sty, style; vb, vascular bundle; xy, xylem.

(d) Parenchyma tissue with cells varying in size from about 15 to 66 microns. Many large stone cells varying from round to elongated occur in this tissue.

(e) Fibrovascular tissue. The vessels vary in size to about 30 microns in diameter, and are arranged in radial rows; scalariform, reticulate, and spiral vessels are present.

(f) Pith, consisting of thick-walled parenchymatous cells with small intercellular spaces, and rosette crystals of calcium oxalate.

Ground Cloves

The most important identifying characteristics of ground cloves are fragments of epidermis tissue, reddish-brown in color; parenchymatous tissue with volatile oil cavities; spiral vessels; thick-walled fibers; crystal fibers; rosette crystals of calcium oxalate; corolla tissue with numerous oil cavities; reticulate cells of anthers; and triangular pollen grains.

As 5 per cent of clove stems are premitted in the ground product it is to be expected that some fragments of the stem tissues will be present. Of these, the most important are the stone cells and the scalariform and reticulate vessels.

Cloves contain no starch. Clove stems contain a very small quantity of starch; the granules are round, about 15 microns in diameter, hilum central, and exhibit a clearly marked cross with polarized light.

VIII. GINGER

Zingiber officinale Roscoe
Family: *Zingiberaceae*

GENERAL

An herbaceous perennial consisting of an underground stem or rhizome which produces erect, leafy shoots 1 to 3 feet high.

Leaves alternate, lanceolate or linear lanceolate, acute, smooth, sessile, sheathing, light green, and up to about 8 inches in length.

Flowers borne in a bracteal, imbricated spike, terminating a leafless stem reaching to about 12 inches in height. Flowers situated in the axils of large, greenish-yellow obtuse bracts, small, few, usually arising one or two at a time yellowish and speckled, with a purplish lip also speckled; evanescent. The fruit is an oblong capsule but it is rarely produced.

The method of harvesting and preparing ginger for market varies in the different countries of cultivation. In general, when the plant is about 9 months old and the green leafy stems commence to turn yellow and cessation of growth is evident the rhizomes are lifted from the earth, cleared of all adhering matter by washing, blanched, and then sun dried. Some producers scald the rhizomes in hot water and then dry them in the sun. Others scald the rhizomes, and then peel or scrape them to remove the cork before drying in the sun. Scraping is thought by some to be a better method than peeling because of the probable loss of aromatic constituents through accidental cutting into the cortex. The rhizomes are usually spread out on some framework to dry slowly in the sun; if spread upon the ground there is always the danger of spotting, blackening, and mold loss developing from the dampness of the earth. It requires about 6 to 8 days to dry the rhizomes sufficiently for market.

In the West Indies it is held that the rhizomes should not contain more than 7 to 12 per cent moisture when ready for market. With a higher moisture content than this, they say, the ginger is liable to ferment. Care is taken not to fracture the rhizomes. Chemical methods of drying are not employed.

Two forms of ginger are exported from Jamaica, the peeled and unpeeled ginger. The peeled ginger is prepared by scalding the rhizomes in hot water and then removing the epidermis with a knife. Jamaica ginger is graded No. 1, No. 2, No. 3, and Ratoon. The last mentioned is an inferior grade of ginger consisting of smaller and more fibrous pieces.

African ginger for export is examined by the produce inspection branch of the Agricultural Department and must conform to the minimum standard of quality laid down by the government of Sierra Leone.

India has three grade classifications for their export ginger, depending upon the tract of production, color, and fiber content of the rhizomes.

MORPHOLOGY

Jamaica Ginger

Rhizomes very light buff in color, and varying in length from about 6 to 9 cm; shape irregular, branched or palmate; divested of cork and epidermal tissue; clean, hard, and somewhat fibrous.

The odor of Jamaica ginger is agreeable, aromatic, and somewhat pungent; the taste is aromatic, pungent, biting.

African Ginger

Rhizomes darker in color than Jamaica ginger, and not completely scraped. Rhizomes complete, dry, plump, with cork removed from the sides only.

African ginger lacks the fine aroma of the Jamaica product but it is very pungent.

Indian Ginger, Cochin Ginger

Rhizomes to about 12 cm in length, cork partly removed on the flattened sides; color, light brown; patches of cork very light brown, wrinkled. **Calicut Ginger:** Rhizomes vary in color from dark yellow to light brown; cork removed from the sides.

The odor of Indian ginger is aromatic, the taste is strongly aromatic and pungent.

Japanese Ginger

Rhizomes resemble those of Cochin ginger in appearance, but the Japanese product is usually limed.

Japanese ginger is weakly aromatic and weakly pungent.

HISTOLOGY

Jamaica Ginger, Scraped

In the preparation of this product the epidermis, hypodermis, and cork are removed, and the spice consists of (a) cortex; (b) endodermis; (c) pericycle; and (d) central cylinder.

Cross section:

Cortex. Consisting of (a) thin-walled parenchyma cells with starch; the starch granules are smallest in the outer cortex tissue; (b) brown-colored oleoresin cells scattered throughout the cortex; and (c) fibrovascular bundles.

Endodermis. An unbroken ring of tangentially elongated cells with thickened, suberized radial walls.

Pericycle. An unbroken ring of tangentially elongated cells situated on the inner side of the endodermis.

Central Cylinder. In the outer cylinder tissue, adjoining the pericycle there is a vascular bundle zone, no fibers. Fibrovascular bundles and oleoresin cells occur in the central region of the cylinder. The ground tissue of the central cylinder consists of thin-walled parenchyma cells containing starch.

The fibrovascular bundles are large. In longitudinal section, the fibers are long with moderately thick walls and wide lumen. The vessels are large and mostly scalariform, excepting in the vascular bundle zone adjoining the pericycle where large reticulate, scalariform, and some spiral vessels occur.

Starch is abundant. The granules are ovate and many are characterized by a protuberance at one end. They vary in size to about 45 microns, long axis, and up to about 24 microns in width at the broad end. With polarized light the granules exhibit a distinct cross through the hilum at the small end.

Unscraped Ginger

In addition to the tissues described above, the following tissues are present in unscraped ginger: (a) epidermis, frequently disorganized, consisting of tangentially oblong-rectangular cells; (b) a hypodermis of a few layers of parenchymatous cells; and (c) several layers of tangentially oblong-rectangular, thin-walled, suberized cork cells.

Ground Ginger

The most important identifying characteristics of ground ginger are (a) the abundance of starch; (b) large vessels; and (c) fibers. The fibers of ginger are only partially lignified, but refractive to polarized light. With chlor-zinc-iodine they stain blue with varying intensity. In samples of *unscraped* ginger there will be seen, in addition to the above, fragments of cork, hypodermis, and epidermal cells.

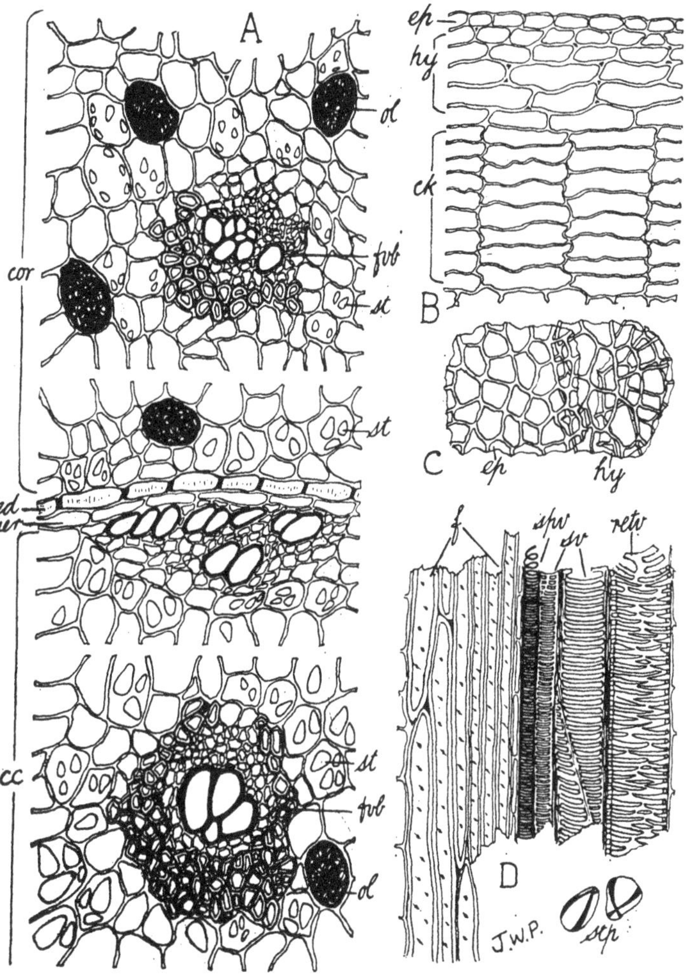

Fig. 8. GINGER. (A) Cross section of Jamaica ginger. (B) Cross section of outer tissues of unscraped ginger. (C) Surface view of epidermis and hypodermis, unscraped ginger. (D) Fibrovascular tissue.

cc, central cylinder; ck, cork; cor, cortex; ed, endodermis; ep, epidermis; f, fibers; fvb, fibrovascular bundle; hy, hypodermis; ol, oleoresin; per, pericycle, retv, reticulated vessel; sv, scalariform vessel; spv, spiral vessel; st starch; stp, starch, polarized light.

IX. MACE

Myristica fragrans Houtt

GENERAL

The mace of commerce is a product of the nutmeg tree. It is the crimson aril which clothes the shell-enclosed nutmeg. The aril is skilfully removed by opening it from the top and turning it back, after which it is flattened out either by hand or between boards. The mace is then placed in the sun for a few hours each day until dry.

Fresh mace is brilliant red in color, and somewhat tough and leathery. Drying changes the color to buff or light brown, and the mace becomes horny and brittle.

MORPHOLOGY

True Mace

This is the aril of the nutmeg seed. It must be differentiated from the aril of *Myristica argentea* War., known as Macassar or Papua mace, and from the aril of *Myristica malabarica* Lam., known as Bombay or Wild mace.

True mace, in the dried, or commercial state, is buff or very light-brown in color. It varies in length to about 4 cm, and consists of flat pieces, branched, smooth, horny and brittle.

The odor of true mace is fragrant and nutmeglike; the taste is aromatic and slightly warm.

Bombay or Wild Mace

This is reddish-brown in color. It is around 4 to 4.5 cm in length, and consists of numerous, narrow branches dividing to form a dense, almost solid, domelike mass of thin, curled branchlets at the apex.

Bombay mace is practically odorless and definitely tasteless. It is an adulterant.

True mace is usually marketed under the name of its place of origin, as, for example, Grenada, Penang, or Banda.

HISTOLOGY

Cross Section of Grenada Mace

In cross section Grenada mace (a true mace) consists of (a) cuticle; (b) an outer epidermis of irregular-shaped cells with thick walls; the outer walls are much thicker than the inner; (c) an hypodermis consisting of a layer of thick-walled cells, and, in places, one or more additional layers, similar in appearance to the epidermal cells. Where additional layers occur, the walls of these cells are much thinner than those of the cells of the main hypodermal layer. It should be observed that the hypodermis is not always a continuous layer, sometimes being broken or interrupted. The cell walls of the epidermis and those of the hypodermis stain pale blue with chlor-zinc-iodine following iodine-iodide solution; (d) a mesophyll consisting of thin-walled parenchyma cells containing amylodextrin granules, densely packed and staining brown with iodine solution; numerous large, oleoresin cells, yellow in color; vascular bundles consisting of xylem and phloem tissues. In longitudinal section,the vessels are mostly scalariform, with some reticulated and spiral vessels present; and, finally, an inner epidermis of thick-walled cells similar to those of the outer epidermis.

In surface view, the epidermis consists of greatly elongated cells with wide walls etched with very fine cross lines; the cells of the hypodermis are polygonal in shape.

Ground Mace

Ground mace in a water mount appears as a pasty mass in the microscope field, but when well rubbed out fragments of the tissues described above are clearly identifiable.

Presence of Bombay Mace

In cross section, the outer wall of the epidermal layer of Bombay mace is thicker than that of true mace. The epidermal cells of Bombay mace tend to radial elongation and lie close together, averaging around 8 cells to 100 microns, whereas those of true mace are irregular in shape and average around 4 or 5 cells to 100 microns. In surface view, the walls and lumens of the epidermal cells of Bombay mace are narrower than those of true mace. A hypodermis is ill-defined or absent in Bombay mace, but present for the most part in true mace. In surface views of fragments of mace consisting of epidermis and underlying tissues the hypodermis is obscure or absent in Bombay mace, but prominent in true mace. In such fragments, the oleoresin cells of Bombay mace will be seen to be more numerous and closely packed than those of true mace. In both products the oleoresin cells are more numerous in the outer mesophyll regions, but more particularly so in Bombay mace. The oleoresin cells of Bombay mace are a bright orange in color, but those of true mace are definitely yellow. On the addition of sodium hydroxide solution the oleoreisin cells of Bombay mace turn red, but the oleoresin cells of true mace remain yellow when similarly treated.

The presence of Bombay mace may be confirmed by the following chemical test. Prepare an alcoholic extract of the mace under examination. To the extract add a few drops of potassium chromate solution. If Bombay mace is present the precipitate and extract will turn red on standing. If Bombay mace is absent the precipitate and extract will be yellow and remain so on standing.

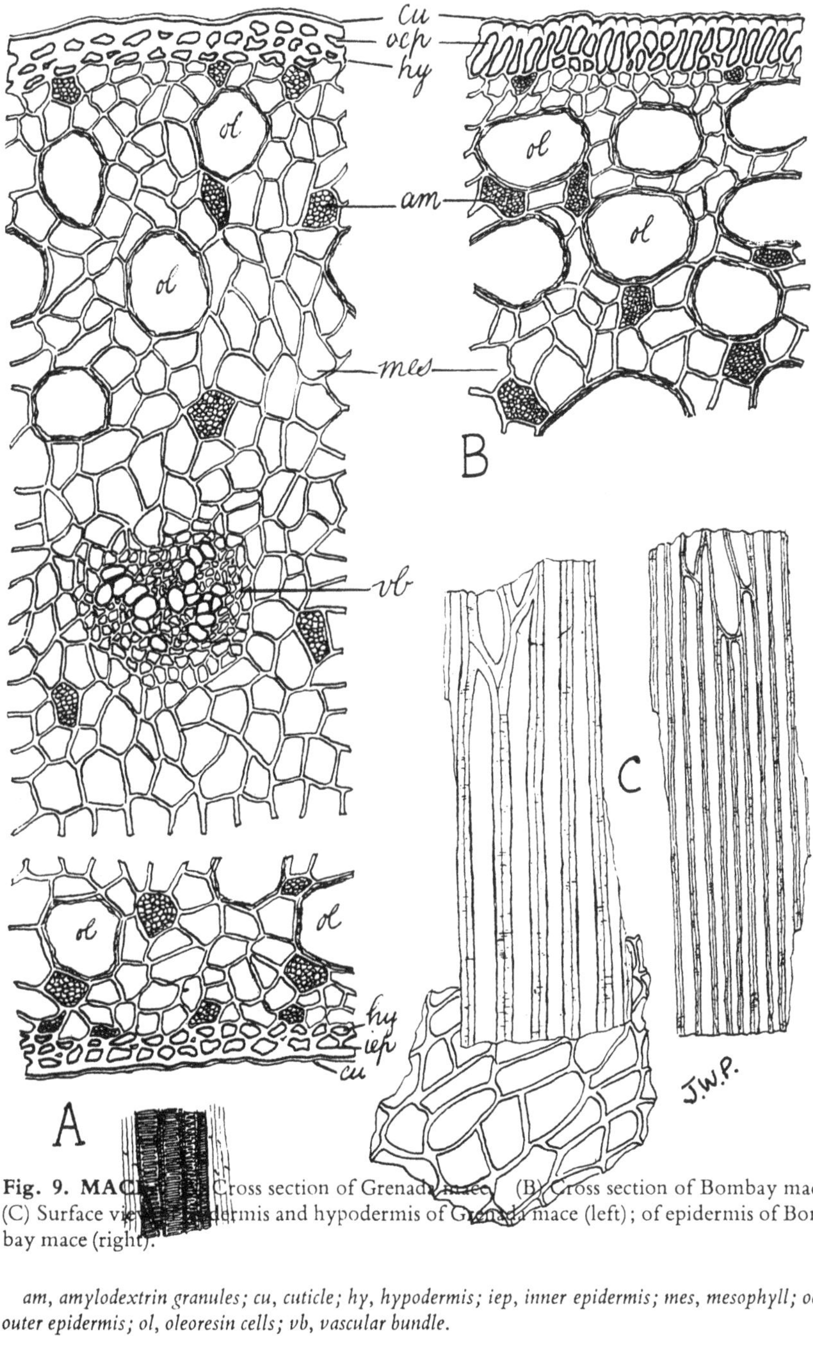

Fig. 9. MA[redacted] Cross section of Grenada mace. (B) Cross section of Bombay mace. (C) Surface vi[redacted]rmis and hypodermis of Grenada mace (left); of epidermis of Bombay mace (righ[redacted].

am, amylodextrin granules; cu, cuticle; hy, hypodermis; iep, inner epidermis; mes, mesophyll; oep, outer epidermis; ol, oleoresin cells; vb, vascular bundle.

X. NUTMEG

Myristica fragrans Houtt
Family: *Myristicaceae*

GENERAL

The nutmeg tree is bushy, and reaches about 30 to 40 feet in height. It is dioecious. The leaves are alternate, oblong-ovate, acute at apex and base, entire, coriaceous; dark green and lustrous on the upper surface, whitish on the lower; petiolate; and about 10 cm, or more, in length.

The flowers are small, fleshy, pale yellow, and aromatic. The staminate (male) and pistillate (female) flowers are borne on separate trees The staminate flowers are larger than the pistillate, axillary, usually in clusters of three.

The pistillate flowers are single, solitary, and axillary.

The fruit when ripe resembles a peach in appearance, and is yellow, smooth, and fleshy. It becomes dry and coriaceous and splits open disclosing a bright red, fleshy, laciniate aril which invests a thin, lustrous, deep brown shell containing a single seed or nutmeg.

In Penang the fruit is sometimes allowed to fall but it is usually collected from the trees with the aid of a bamboo stick about three feet long, split several times close to one end and expanded to form a bulbous cavity surmounted by a two-pronged projection. The nutmeg kernels are dried in the sun, and when, on shaking, the nutmegs rattle in the shells they are ready for storing. The nutmegs are later freed from the shells by cracking them open with a wooden truncheon, or by mechanical means.

In Grenada, British West Indies, the fruits are allowed to fall and it is considered unsound to "reap" them from the trees. The freshly fallen fruits are gathered each morning and carried to "boucans" or curing houses for processing. It requires from 6 to 8 weeks to cure the nutmegs in the air and sun.

Nutmegs are usually graded by weight, the three common grades averaging

65, 85, and 110 nutmegs per pound respectively, but are also exported unassorted.

MORPHOLOGY

Nutmegs are grayish-brown in color; mostly ovoid or ellipsoidal but some nearly globular in shape, and varying in size. The large ovoid nutmegs are around 30 mm in length and 20 mm in thickness. The surface is furrowed, wrinkled, with a raphe depression or groove extending from a scar at the base of the seed to a depression at the apex. Nutmegs are quite hard, yet cut easily. The cut surface presents a pale brown color (endosperm) marked by many brown veins (perisperm). It is in these veins that the essential oil of nutmeg is found, and pressure upon any one of them will cause an exudation of oil. When halved longitudinally the remains of the embryo are discernible.

Nutmegs have a characteristically aromatic odor, and an aromatic, warm, slightly bitter taste.

HISTOLOGY

In cross section, nutmeg consists of (a) an outer perisperm of tangentially elongated, thick-walled cells filled with reddish-brown pigment and containing prismatic crystals, probably acid potassium tartrate; small intercellular spaces; (b) an inner perisperm with arms branching into the endosperm, and consisting of isodiametric cells with reddish-brown contents, and, in those parts branching into the endosperm, numerous large oil cells; (c) vascular bundles of phloem and xylem tissue situated in the inner perisperm. The vessels are spiral; (d) a layer of compressed cells separating the inner perisperm from the endosperm; (e) an endosperm of polygonal parenchyma cells. Starch is the chief constituent of the endosperm cells. The starch granules are round or truncated, with central hilum, and from about 3 to 22 microns in diameter; with polarized light they exhibit a distinct cross. The endosperm appears whitish in the microscope field with occasional yellow-orange pigmented areas.

In surface view, the cells of the outer perisperm are isodiametric in shape. The crystals are clearly visible with ordinary transmitted light, and refractive to polarized light.

Embryo

The remains of the embryo may be removed from the seed by soaking the nutmeg in water and removing the surrounding tissues from the radicle and branched cotyledons. Noteworthy features of the embryo are the numerous oil cells and long, canal-like secretion cells, which, in the cotyledons, lie very close to, and almost surround, the vascular tissues. In the radicle there is similar structure but, in addition, there are numerous oil cells scattered throughout the parenchyma tissue. The vascular tissue is well developed, and spiral, reticulated, and scalariform vessels will be observed in the xylem.

Ground Nutmeg

Because of the large quantity of fixed oil in nutmeg, the ground sample in a water mount appears as a thick, pasty mass in the microscope field. It must be well rubbed out. Fragments of endosperm are predominant; other tissues to lesser degree. The starch is abundant and significant, being easily identified with ordinary transmitted light, and conspicuous with polarized light. The starch is seen in single granules, in groups of 2, 3 or 4 granules, and in large aggregates.

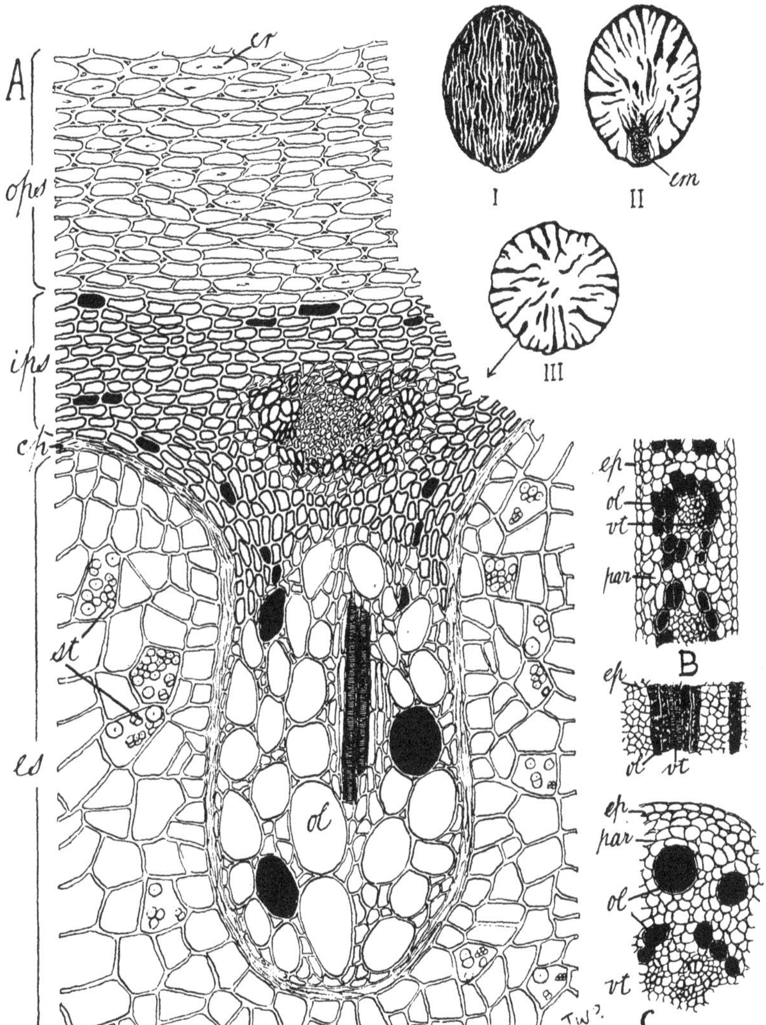

Fig. 10. NUTMEG. I—Whole nutmeg. II—Nutmeg, longitudinal section. III—Nutmeg, cross section. (A) Cross section through perisperm and endosperm. (B) Cotyledon, cross section (above); surface view (below). (C) Cross section of a portion of radicle.

cp, compressed parenchyma; cr, crystals; em, embryo; ep, epidermis; es, endosperm; ips, inner perisperm; ol, oleoresin cells; ops, outer perisperm; par, parenchyma; st, starch; vt, vascular tissue.

XI. PEPPER

Piper nigrum L.
Family: *Piperaceae*

GENERAL

The pepper plant is a perennial, climbing shrub or vine with a smooth, woody, articulate stem swollen at the joints. The vine clings to a tree or other support by means of numerous short rootlets produced at the joints of the stem. In its natural state the pepper vine may reach a height of 20 to 25 feet, but under cultivation it is usually kept down to 15 feet.

The leaves are alternate, broad-ovate, accuminate, coriaceous, smooth, petiolate, and dark green.

The flowers are very small, whitish, sessile, and borne in pendulous, dense catkin-like spikes.

The fruit is small, berry-like, containing a single seed, sessile, nearly globular; at first green, then yellowish, and finally red when ripe.

When the berries begin to turn red the spikes are gathered from the pepper vines. The berries are removed from the spikes by hand or by beating with short sticks. The stalks or stems are separated from the berries by winnowing. The quality of the berries is said to be improved if they are next plunged into boiling water. Ten minutes of this treatment has the effect of turning the berries black in an hour. Drying is then carried out by spreading the berries on mats in the full sun, and is completed in about 3 to 4 days.

White pepper is obtained by depriving the pepper berries of the greater part of the pericarp, or outer coating as it is commonly called. This is accomplished by keeping the ripe berries for 2 or 3 days in moist heaps, or in sacks submerged in running water. This process softens the dark-colored pericarp which is then almost entirely removed by washing the berries in baskets, or by trampling them in a tub or tank followed by washing. The white berries are then spread out on mats and allowed to dry in the sun.

MORPHOLOGY

Black Pepper

Its berries or peppercorns are dark brown to black in color; nearly globular in shape; around 5 mm or less in diameter; with deep-set wrinkles forming a characteristic network. The pericarp is thin and encloses a single seed with a hollow center. The perisperm of the seed is horny in the outer part, floury around the central cavity. The embryo is embedded in a small endosperm at the apex of the seed.

Black pepper has a characteristic, penetrating, aromatic odor; and a hot, biting, very pungent taste.

White Pepper

Its berries are light yellowish-gray in color; nearly globular in shape; around 5 mm or less in diameter; smooth; striated; with a small protuberance at the base and another arising from a depression at the apex. The striations are very distinct on some berries, indistinct or not visible on others; they are vascular bundles.

The odor and taste of white pepper is similar to that of black pepper but less pungent.

HISTOLOGY

In cross section, the black pepper berry consists of the following tissues:

Pericarp

This consists of: (a) an exocarp of nearly rectangular cells and a well defined cuticle. In surface view the cells of the exocarp are polygonal, darkly pigmented, and contain small prismatic crystals. Occasional stomata will be observed; (b) hypodermis of stone cells interspersed in places with thin-walled polygonal cells. The stone cells of this supporting tissue are varied in shape but the greater number are radially elongated; (c) an outer mesocarp of moderately thick-walled, irregularly-shaped parenchyma cells, and scattered oleoresin cells; (d) a vascular bundle zone; (e) an inner mesocarp of sclerenchymatous cells irregular to tangentially elliptical in shape. These cells are smaller and the walls are thicker than those of the outer mesocarp; (f) a zone of sclerenchymatous cells varying in shape from isodiametric to tangentially oval or elliptical, and in size to about 85 microns, long axis, containing oil drops; (g) a layer of sclerenchymatous, porous (beaded) cells; and (h) an endocarp consisting of a layer of sclerenchymatous cells with thickened radial walls and very thick inner walls, commonly called "beaker cells."

At the base of the berry the walls of the various tissues are generally much thicker than elsewhere in the fruit; many walls are irregularly thickened, some are beaded. Stone cells are present. Here, the endocarp beaker cells modify and converge with other tissues, the inner cuticles terminate, and the vascular tissues enter the fruit.

The xylem tissues of the berry consist of xylem parenchyma, annular, and spiral vessels.

With chlor-zinc-iodine following bleaching, the cuticle, hypodermis, and all tissues from the vascular bundle zone to the endocarp inclusive stain yellow, indicating, with the exception of the cuticle, their sclerenchymatous nature (cuticle always stains yellow with chlor-zinc-iodine). If the section is left too long in the bleaching solution the lignin in the cell walls of the latter group of tissues will dissolve and the cell walls will stain blue (cellulose).

Seed

The tissues of the seed are (a) a thin cuticle; (b) seed coat consisting of an outer layer, connective layer, and inner layer of tangentially oblong or elliptical cells. Longitudinally, the cells of the outer layer are oblong; those of the connective layer, polygonal; and those of the inner layer, elongated; the latter with beaded walls. The cells of the outer and inner layers are darkly pigmented. At the apex of the seed the cells of the outer layer become isodiametric in shape, those of the connective layer are disintegrated, and those of the inner layer become palisade in shape. At the base of the seed, the seed coat thickens into a mass of darkly pigmented cells about 75 microns deep; (c) a thick cuticle which wrinkles severely in the apex region; and (d) the perisperm, consisting of an outer layer of polygonal cells containing aleurone grains, and inner layers of polygonal cells, varying in size to about 110 microns, long axis, containing masses of starch granules. Occasional oleoresin cells occur in the perisperm. Prismatic crystals occur scattered in the perisperm.

In the seed, the cuticles stain yellow, the seed coat blue, the aleurone layer brown and the starchy perisperm blue.

Endosperm and Embryo

The endosperm and embryo must be studied in longitudinal section. They are usually seen as a disorganized mass of tissue, but repeated sectioning and prolonged study will bring to view the small, polygonal parenchyma cells of the endosperm and the inner parenchyma cells of the embryo. These are not significant in a microscopical examination of ground pepper.

Ground Pepper

Fragments of the tissues just described will be observed in the microscope field of ground black pepper. The presence of blackish-brown fragments of the exocarp, and the stone cells of the hypodermis distinguish black from white pepper. In the fields of both peppers the sclerenchymatized beaker cells of the endocarp and the large masses of starch are prominent. Pepper starch is seen in the microscope field as irregular-shaped masses, with individual granules scattered about. Individual granules are very small, varying in size from about 2 to 6 microns with the majority of granules being from 3 to 4 microns. The shape is polyhedral; the hilum is central and visible with the high power of the microscope and subdued, transmitted light. With high power and polarized light a cross is faintly visible.

Related Pepper

Long Pepper. Long pepper is the compound fruit of two species of the pepper family—namely, *Piper officinarum* C. DC. and *Piper longum* L. The former is cultivated chiefly in Java, the latter in Bengal. The fruits of both species are elongated, cylindrical, catkin-like spikes of immature individual fruits fused together in spirals. The compound fruits vary in size from about 2 to 6 cm in length, and from about 4 to 8 mm in breadth. The color is grayish-black masking a reddish-brown visible after washing. The odor is aromatic; the taste is aromatic and pungent.

The principal histological features of the pericarp are an epicarp of polygonal-shaped cells similar to those of the epicarp of black pepper; a hypodermis with elongated stone cells; a mesocarp with a fibro-vascular bundle zone, volatile oil secretion cells, and reticulated palisade-like cells adjoining the endocarp; an endocarp of elongated, porous, sclerenchymatous cells. Because the small, individual fruits are coalesced, the epicarp and hypodermis exist only in the outer side of the fruit. The chief features of the seed are the sclerenchymatous outer layer of the seed coat, the elongated cells of the perisperm, and the presence of starch grains larger than those of black pepper.

Long pepper was employed as a spice from ancient to early modern times in Europe when its use rapidly dwindled. It is not important today and is probably employed more in southeast Asia than other parts of the world.

Cubebs. The cubeb is the fruit of *Piper cubeba* L. fil. It is a nearly globular fruit similar in size to black pepper. The base of the pericarp is extended into a pedicel-like formation. The surface is coarsely reticulated and brownish in color. Usually the seed is free from the pericarp except for a weak attachment at the base of the pericarp or, not infrequently, to some part of the pericarp other than the base. The odor is aromatic; the taste is aromatic and pungent

The principal histological features of the pericarp are an epicarp of thick, wavy-walled cells; an hypodermis of rectangular or polygonal, porous, thick-walled stone cells with intervening parenchyma tissue; an outer mesocarp of parenchyma tissue and numerous volatile oil secretion cells; an inner mesocarp of tangential, somewhat compressed parenchyma cells traversed by fibrovascular tissues; and an endocarp of isodiametric to radially oblong stone cells with thick, porous walls. The chief features of the seed are the parenchymatous seed coat; a perisperm of polygonal-shaped cells; a small endosperm; and a minute embryo. The individual starch grains are larger than those of black pepper.

Cubebs find some use in medicine but none in the spice trade.

African Black Pepper, Nigerian Black Pepper, African Cubebs. This pepper is the dried fruit of *Piper clusii* C. DC. The fruit is nearly globular with the base of the pericarp extended into a pedicel-like structure up to about 8 mm in length. The surface is rough, and the color varies from light to dark-brown. The size of the fruits vary from about 2 to 5 mm in diameter. The seed is firmly attached to the pericarp. The odor is aromatic; the taste is aromatic and pungent. Many pieces of stem are present, some with fruits attached, in a sample of this pepper.

The principal histological features of the pericarp are an epicarp of small, polygonal-shaped cells with occasional stomata; an outer mesocarp of parenchyma tissue in which many oleoresin cells occur; a central mesocarp of about seven layers of tangentially elongated parenchyma cells traversed by vascular tissues; an inner mesocarp of parenchymatous tissue with numerous large oil cells; and a parenchymatous endocarp. There are no stone cells in the pericarp. The chief features of the seed are the seed coat of three layers of compressed parenchyma cells with the outer walls of the outer layer being thickly cutinized; a perisperm of poygonal-shaped cells containing masses of starch; a small, parenchymatous endosperm in which is embedded a minute embryo composed of tiny parenchyma cells with a number of small oil cells occurring toward the radicle.

The pedicel-like structure at the base of the pericarp consists of an epidermis, a parenchymatous cortex in which many oleoresin cells occur in the outer region, a complete ring of fibro-vascular tissue, and a central region of thick-walled parenchyma cells.

The stem consists of an epidermis of very small cells with many multicel-

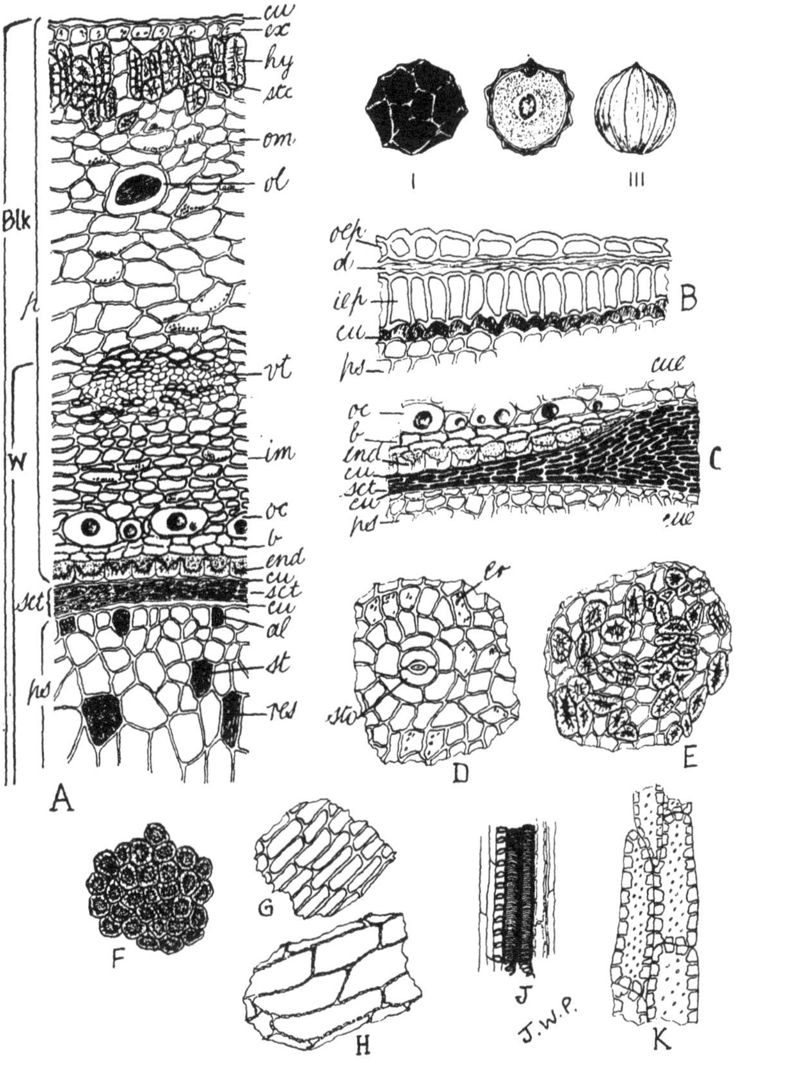

Fig. 11. PEPPER. I—Black pepper. II—Black pepper, cross section. III—White pepper. (A) Pericarp, seed coat, and portion of perisperm; Blk, black pepper; W white pepper. (B) Seed coat in apex region. (C) Seed coat in base region. [A,B, and C, in cross section.] (D) Exocarp. (E) Hypodermis. (F) Endocarp. (G) Outer epidermis of seed coat. (H) Inner epidermis (inner layer) of seed coat. [D, E, F, G, and H, in surface view.] (J) Vascular tissue. (K) Sclerenchymatous cells occurring in base region of pericarp.

al, aleurone; b, beaded cells; cr, crystals; cu, cuticle; cue, cuticle ends (terminates); d, disintegrated cells; end, endocarp; ex, exocarp; hy, hypodermis; iep, inner epidermis (inner layer of seed coat); im, inner mesocarp; oc, oil cells; oep, outer epidermis; ol, oleoresin; om, outer mesocarp; p, pericarp; ps, perisperm; res, resin; sct, seed coat; st, starch; stc, stone cells; sto, stoma; vt, vascular tissue.

lular, thin-walled, pointed hairs; a parenchymatous cortex; and a ring of fibro-vasuclar tissues. The central (pith) region is hollow.

This pepper is employed in Africa as a spice.

XII. SAFFRON

Crocus sativus L.
Family: *Iridaceae*

GENERAL

A perennial, low-growing plant with an underground globular corm. Leaves radical, sheathing, sessile, linear, dark green on the upper surface with a central, longitudinal, white line or furrow, paler on the under surface.

Flowers borne on a scape, blooming in the autumn. Perianth pale reddish-purple, with a cylindrical tube and 6 oblong-oval segments. Stamens 3; ovary inferior and 3-loculed. Style slender, elongated, pale yellow, and terminating in 3 drooping, deep red stigmas.

Only the stigmas are employed in the spice trade.

MORPHOLOGY

Dried stigmas, 3 in number surmounting a short length of style, or free. Stigmas dark red, convoluted, trumpet-shaped, scalloped or notched at the distal end, usually about 25 mm long. Style yellowish.

Odor strong, suggestive of iodine, unpleasant, aromatic; taste aromatic, bitter.

HISTOLOGY

The cells of the stigmas are elongated, narrow, thin-walled, and contain coloring matter. Short papillae are a characteristic of the epidermal cells. Long papillae, varying in length to 150 microns and in width to about 45 microns, occur at the distal ends of the stigmas. Phloem, and xylem tissue with delicate annular and spiral vessels complete the tissue.

Spherical pollen grains up to about 120 microns in diameter are usually present among the long papillae.

XIII. MEXICAN OR AMERICAN SAFFRON

Carthamus tinctorius L.
Family: *Compositae*

GENERAL

An annual herb grown in the temperate regions of the central Mexican tablelands. The plant reaches about 1 foot or more in height. It has an erect stem, somewhat branched at the top, ovate-oblong leaves, and orange-colored flowers.

MORPHOLOGY

Tubular florets, deep orange-red in color, from about 2.5 to 4 cm in length; corolla gamopetalous, cylindrical, with 5 narrow lobes; stamens 5, arising from the corolla tube, anthers encircling the style; style bilobed, yellow.

Odor strong, unpleasant, sweetish and aromatic; taste aromatic, bitter.

HISTOLOGY

Corolla

In surface view, the epidermal cells of the corolla tube are nearly oblong-rectangular in shape and straight-walled. Those of the corolla lobes are elongated and have sinuous walls, except at the apexes where the cells are small and papillose. A few papillae of moderate length occur on the outer epidermis near the apexes of the lobes. Numerous prismatic crystals of calcium oxalate are present in the apex cells. Five vascular bundles and 5 oil canals traverse the mesophyll of the corolla tube. Where the lobes begin these bundles and canals branch and traverse the marginal mesophyll of the lobes.

Stamens

In surface view, the epidermal cells of the filament are small, thin-walled and oblong-rectangular. A few multicellular, multiseriate hairs occur near the juncture of filament and anther. A single vascular bundle containing delicate spiral vessels traverses the filament.

The epidermal cells toward the base of the anther are oblong-rectangular to nearly quadrilateral in shape and have strong but not thick walls. Toward the apex of the anther the cells bordering the dehiscent tissue and extending to the margins and apex of the anther are thick-walled and porous. In the region of dehiscence the cells are oblong-rectangular and thin-walled. The pollen grains are spherical, spinose, and up to 50 microns in diameter.

Style and Stigma

In surface view, the epidermal cells of the style are oblong-rectangular and thin-walled. A fibrovascular bundle and 2 oil canals traverse the style; they branch and traverse the lobes of the stigma. The fibers are long, narrow, and of cellulose.

The stigmatic surface is bristling with spinose papillae up to 180 microns long at the bulbous base of the stigma and becoming progressively shorter toward the apexes of the lobes; at the apexes the papillae are very small. Pollen grains are usually present among the papillae, many germinated and exhibiting pollen tubes.

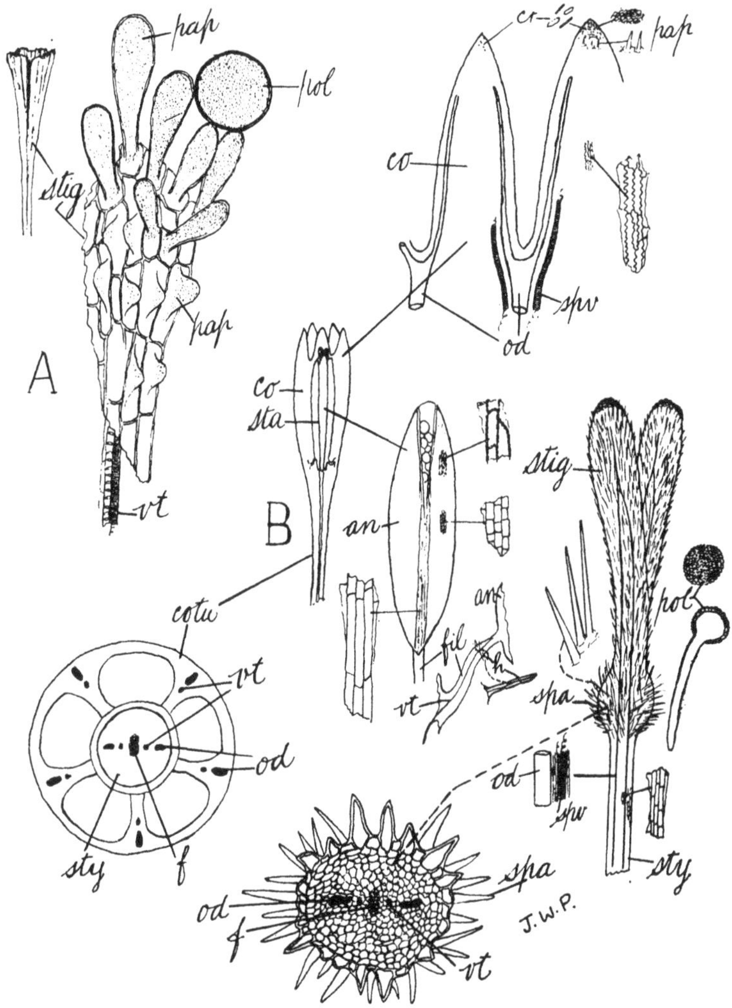

Fig. 12. SAFFRON. (A) True saffron. (B) Mexican saffron (safflower).

an, anther; co, corolla; cotu, corolla tube; cr, crystals; f, fibers; fil, filament; h, hair; od, oil duct; pap, papilla; pol, pollen; spa, spinose papilla; spv, spiral vessel; sta, stamen; stig, stigma; sty, style; vt, vascular tissue.

XIV. STAR ANISE

Illicium verum Hook
Family: *Magnoliaceae*

GENERAL

An evergreen tree reaching a height of about 18 feet. Leaves elliptical to oblanceolate, glandular-punctate, entire, acute. Flowers greenish-yellow, axillary. Fruit consisting of a number of woody follicles or carpels arranged around a central axis, each carpel containing a single seed.

MORPHOLOGY

Fruit whorl from 2.5 to 4.5 cm in diameter, individual carpels from about 9 to 19 mm in length; surmounted on a curved peduncle; carpels laterally broad and flat where they join the central axis, bellied and tapered to a point; boat-shaped. Carpels hard, woody, wrinkled, rough, rigid, reddish-brown; internally smooth, lustrous and light brown. Seed light brown, compressed ovoid, about 8 to 9 mm long, 6 mm broad; smooth, lustrous, with conspicuous hilum scar and raphe ridge. Seed coat brittle; endosperm bulky; embryo disorganized.

The carpels of star anise have a pleasant, anise-like odor, and an agreeable, aromatic, sweet, anise-like taste; the seeds to a lesser extent.

HISTOLOGY

Pericarp

Cross section.

Exocarp. Consisting of cuticle and a layer of epidermal cells with thick outer walls, up to 20 microns in places.

In surface view, the cuticle is striated. The epidermal cells are polygonal in shape, and vary in size to about 133 microns, long axis. Stomata are present but not numerous.

Mesocarp. Consisting of parenchyma cells, isodiametric in shape or nearly so, with brown contents. The cells increase in size toward the central zone of the mesocarp, where they reach about 220 microns, and decrease toward the endocarp. The walls of the inner mesocarp cells are thicker than those of the central mesocarp cells, increase in thickness toward the dehiscent side of the carpel, and in the dehiscent zone adjoining the endocarp the cells pass into lignified fibers; in longitudinal section these fibers are long and pitted.

Resin cells and irregular-shaped stone cells are scattered throughout the mesocarp. The resin cells vary in size to about 220 microns, long axis, and contain yellow to brownish-yellow oleoresin. The stone cells vary in size and thickness and are noteworthy for their branching and irregularity.

Vascular bundles occur in the merging zone of the central and inner mesocarp, and consist of phloem tissue and spiral, scalariform, and scalariform-reticulate vessels.

Endocarp. Consisting of a layer of thin-walled, sclerenchymatous palisade cells up to about 440 microns, long axis, decreasing in size toward, and becoming stone cells in the dehiscent region.

In surface view, the thin-walled cells are polygonal in shape, and vary in size to about 120 microns, long axis, most of the cells being around 90 to 100 microns.

Seed

Cross section.

Seed Coat. Consisting of (a) an epidermis of thick-walled, sclerenchymatous, pitted palisade cells up to about 200 microns, long axis; (b) 5 layers of sclerenchyma cells; and (c) 2 or more layers of thin-walled parenchyma cells containing numerous prismatic crystals of calcium oxalate.

In surface view, the epidermal cells are up to about 66 microns, long axis, and show considerable thickening with branching pits. The sclerenchyma tissue can be divided into 3 layers of irregular-shaped cells on the outer side, and 2 layers of elongated, narrow cells on the inner.

Endosperm. Consisting of polygonal cells varying in size to about 110 microns, long axis, and containing aleurone grains and globules of fixed oil.

The aleurone grains are large and each contains a correspondingly large crystalloid; globoids also occur.

Embryo. The tissue is disorganized.

Peduncle

In cross section consisting of (a) striated cuticle; (b) a layer of epidermal cells with thick outer walls; (c) cortex consisting of parenchyma cells varying in size to about 155 microns, long axis, oleoresin cells to about 90 microns, and branching stone cells varying greatly in shape and size; (d) a ring of fibrovascular tissue; and (e) pith with isodiametric cells varying in size to about 155 microns.

Receptacle

The receptacle or central axis is anatomically similar to the peduncle but branching stone cells, and pitted cells, are more abundant.

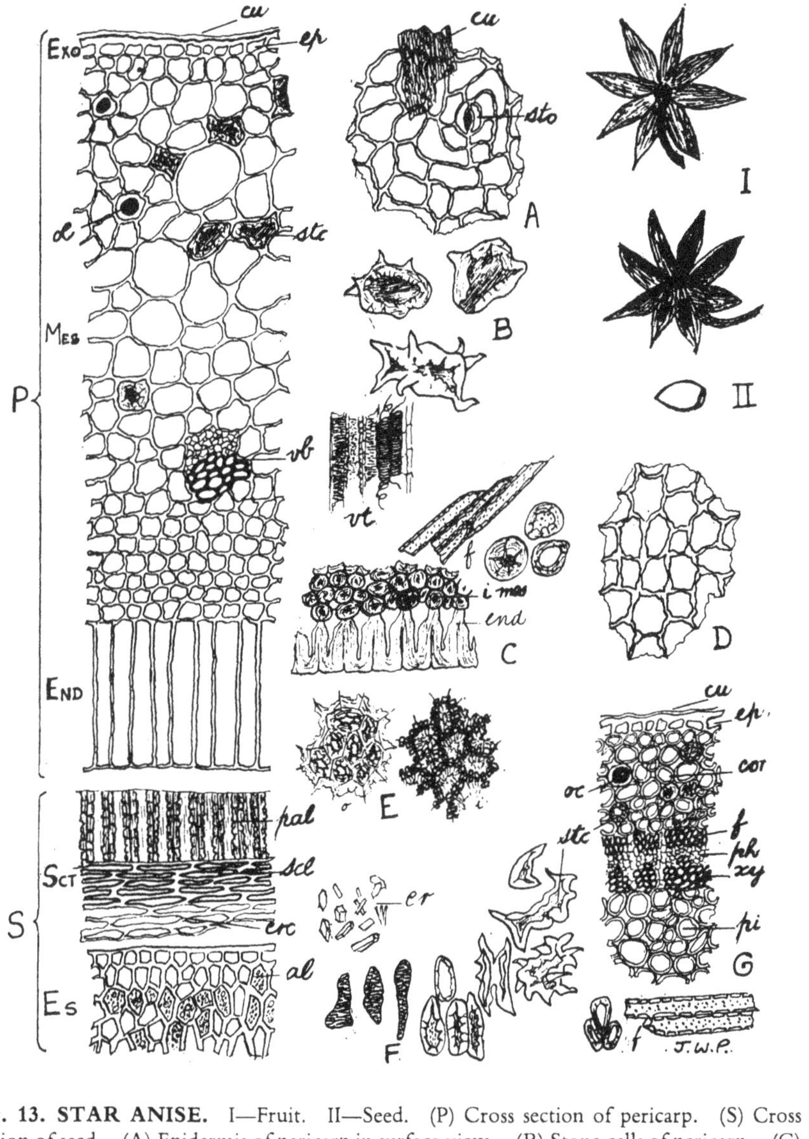

Fig. 13. STAR ANISE. I—Fruit. II—Seed. (P) Cross section of pericarp. (S) Cross section of seed. (A) Epidermis of pericarp in surface view. (B) Stone cells of pericarp. (C) Inner mesocarp and endocarp at dehiscent edge of carpel. (D) Endocarp in surface view. (E) Outer and inner (o and i) surface views of palisade cells of seed coat. (F) Reticulated cells and stone cells occurring in the region of the raphe. (G) Cross section of peduncle.

Al, aleurone; cor, cortex; cr, crystals; crc, crystal cells; cu, cuticle; end, endocarp; ep, epidermis; es, endosperm; exo, exocarp; f, fibers; i.mes, innermesocarp; mes, mesocarp; oc, oil cell; ol, oleoresin; pal, palisade cells; ph, phloem; pi, pith; scl, sclerenchyma cells; sct, seed coat; stc, stone cells; sto, stoma; vb, vascular bundle; vt, vascular tissue; xy, xylem.

XIV. TURMERIC

Genus: *Curcuma longa* L.
Family: *Zingberaceae*

GENERAL

An herbaceous plant with a perennial rootstock or rhizome from which arises tufts of large, broad, lanceolate, bright green leaves, acute at both ends, reaching 2 to 3 feet in height.

Flowers yellow or pale-yellow, borne in a spike terminating a stem shorter than the leaves. The flowers arise from 2 buds situated in the axils of bracts, and mature successively. Bracts are greenish-white, the uppermost tinged with pink.

When the rhizomes are harvested they are first washed to free them from adhering earth, and then cured by placing them in earthenware or metal pots to which is added a few of the dried leaves of the turmeric plant and just enough water to cover the rhizomes. The pots are covered and placed over a slow fire for about three hours and then allowed to cool. Following this process, the rhizomes are spread in the sun to dry for 5 to 7 days. It is customary to polish the dried rhizomes in some manner, usually by rotating them in a horizontally mounted cylinder for a period of ten minutes.

MORPHOLOGY

Alleppey Turmeric

The rhizomes are deep-yellow to orange-yellow in color, and from about 2.5 to 7.5 cm in length, from about 7 to 15 mm in thickness; almost cylindrical, curved, and tapering at the ends; palmate (known as fingers); with tubercle-like protuberances; or plainly cylindrical but thicker toward one end (known as bulbs). Rhizomes are rough, hard, and with numerous encircling ridge-like rings or annulations. When fractured, the break is clean, not splintery or fibrous; surface waxy or horny and resinous in appearance. On the fractured surface the endodermis is clearly visible separating the cortex from the central cylinder.

Turmeric has a characteristic pepperlike odor, and a slightly aromatic, somewhat bitter taste.

Madras Turmeric

The rhizomes are mustard-yellow in color. Similar in other respects to the Alleppey product.

Haiti Turmeric

The rhizomes are a dull yellowish-brown in color; mostly small; of poor appearance; product usually includes broken pieces and longitudinally split rhizomes; otherwise similar to the Indian turmeric.

HISTOLOGY

Cross section of rhizome:

Epidermis

This consists of a layer of tangentially oblong-rectangular cells with occasional unicellular, nonglandular, thick-walled hairs.

Hypodermis

This consists of two or three layers of irregular-shaped parenchymatous cells.

Cork

This consists of about six layers of tangentially oblong-rectangular cells arranged in orderly radial rows.

Cortex

This consists of (a) thin-walled parenchyma cells containing starch; (b) oleoresin cells; and (c) vascular bundles.

Endodermis

This is a layer of tangentially oblong-rectangular cells with thickened radial walls; globules of oil are present in these cells.

Pericycle

This is a layer of tangentially oblong-rectangular cells adjoining the endodermis on the inner side.

Central Cylinder

This consists of (a) vascular bundle zone adjoining the pericycle; (b) parenchyma cells containing starch; (c) oleoresin cells; and (d) vascular bundles scattered in the central tissue region.

The vascular tissue of turmeric consists of phloem and xylem; there are no fibers. The vessels are spiral or scalariform.

Epidermis in Surface View

The cells of the epidermis are polygonal in shape, vary in size to about 120 microns, long axis, and have moderately thick, beaded walls. The beading is best observed in cleared tissue with powers from $\times 250$ to $\times 500$. The epidermis may be cleared with the aid of chloral hydrate, or decolored with bleaching solution. It should then be treated with nitric acid to dissolve the crystalline matter in the cells. Very small prismatic crystals, and groups or masses of crystals without definite shape occur in the epidermal cells.

A few stomata occur in some parts of the epidermis. They are about 30 microns, long axis, with stomatal opening from about 18 to 22 microns. Each stoma is surrounded by 4 epidermal cells.

The hairs are mostly straight, pointed, unicellular, with moderately thick walls. They vary in length to about 600 microns, and in basal width from about 30 to 50 microns. The hairs are fairly numerous but they are not distributed evenly, more occurring in some areas than in others (as with the stomata).

Starch

During the scalding or boiling process used in the preparation of this spice for market almost all of the starch granules are partially cooked leaving them without uniform characteristics. The granules may be whole or fragmented; they are irregular in shape, and they exhibit degrees of shell-like markings. Undamaged starch granules are similar to those of ginger. Turmeric starch is usually seen in the microscope field as irregular-shaped masses or lumps stained yellow with curcumin. On the addition of a drop of iodine test solution these lumps turn dark blue, identifying them as starch.

Ground Turmeric

The most important identifying characteristics of ground turmeric are

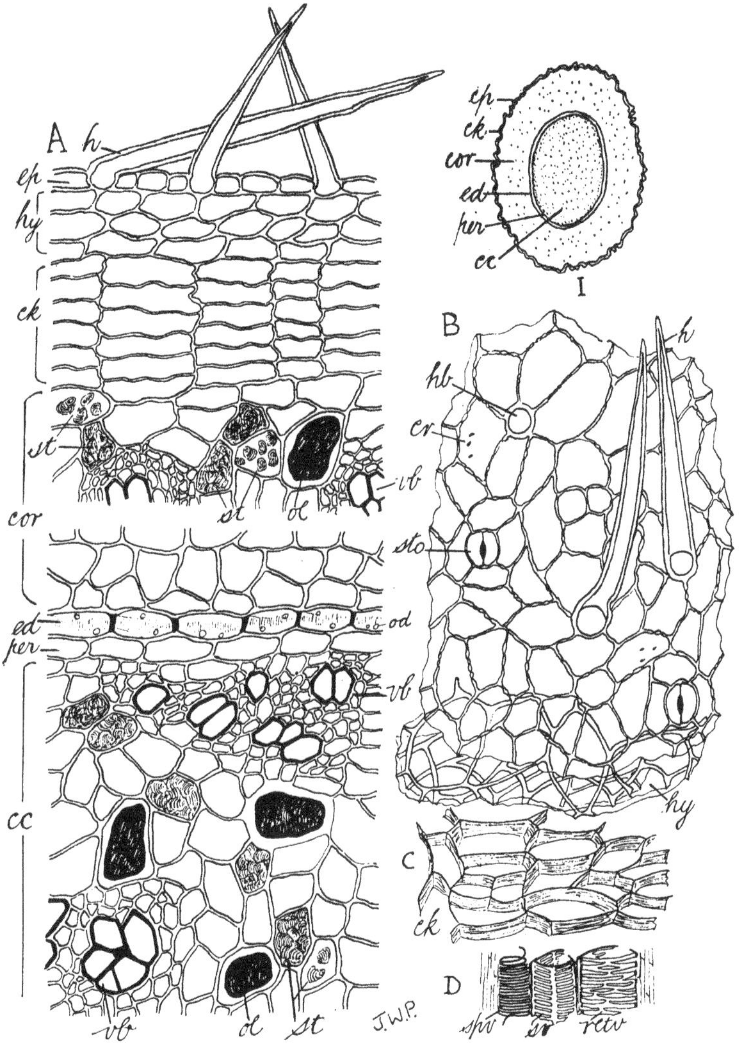

Fig. 14. TURMERIC. I—Whole rhizome, cross section. (A) Cross section to central cylinder of rhizome. (B) Surface view of epidermis and hypodermis. (C) Surface view of cork cells.—(D) Vascular tissue.

cc, central cylinder; ck, cork; cor, cortex; cr, crystals; ed, endodermis; ep, epidermis; h, hair; hb, hair base; hy, hypodermis; od, oil drops; ol, oleoresin; per, pericycle; retv, reticulated vessel; sv, scalariform vessel; spv, spiral vessel; st, starch; sto, stoma; vb, vascular bundle.

(a) numerous irregular-shaped, shell-like, yellow starch lumps; (b) oleoresin lumps, which appear dark, smooth and shiny; (c) fragments of cork cells, and abundant parenchyma tissue; and (d) large, spiral and scalariform vessels of the xylem.

Miscellaneous Seeds

XVI. FENUGREEK

Trigonella foenum-graecum L.
Family: *Leguminosae*

GENERAL

This is an annual herb reaching a height of about 2 feet. Leaves light green, pinnately trifoliate; leaflets obovate. Flowers papilionaceous, small, white. Fruit a legume, long, narrow, curved, tapering into a slender point, and containing small, deeply furrowed seeds.

MORPHOLOGY

Seed yellowish-brown, 2.5 to 5 mm long, 2 to 3.5 mm broad; oblong-rhomboidal, with a deep furrow running obliquely from one side dividing the seed into a larger and smaller part; smooth, and very hard, hilum in the notch, short raphe; translucent endosperm; yellow embryo, with cotyledons occupying the larger part of the seed and the radicle the smaller. Seed becomes coated with mucilage when soaked in water.

Fenugreek has a strong, pleasant, sweetish odor reminiscent of that of burnt sugar. The taste is decidedly farinaceous and slightly bitter.

HISTOLOGY

Seed

Cross section.

Seed Coat. This consists of: (a) cuticle; (b) a layer of thick-walled columnar or palisade cells, flat at the base, mostly pointed but some flattened at the apex, from about 65 to 70 microns high, and 8 to 20 microns wide, with conspicuous light line about 22 microns from the outer end; (c) a layer of supporting or bearer cells, low, tangentially wide, with radial rib-like thickening; and (d) 4 or 5 layers of tangentially elongated, thin-walled parenchyma cells.

In surface view, with low power, the palisade layer appears as a mass of very small, papillose-like cells; with higher power, focused down, the cells are polygonal with pitted thickening. The bearer cells are polygonal but the thickening causes them to appear isodiametric or elliptical; they vary in size up to about 85 microns, diameter or long axis.

Endosperm. This consists of: (a) a layer of thick-walled cells containing aleurone grains; and (b) several layers of thin-walled, mucilaginous cells varying in size to about 250 microns, long axis, radially elongated in the outer, and tangentially elongated in the inner endosperm.

In surface view, the aleurone cells are polygonal in shape, and vary in size up to about 45 microns, long axis.

Embryo. In cross section, the radicle consists of a layer of small epidermal cells; several layers of isodiametric, or nearly so, parenchyma cells varying in size up to about 50 microns, and a central core or zone of very small isodiametric cells varying in size to about 18 microns. The cotyledons consist of (a) an upper and lower epidermis of small cells; (b) mesophyll of 3 to 4 layers of palisade cells varying in size up to about 80 microns, long axis, and several layers of rudimentary spongy tissue; and (c) rudimentary vascular tissue situated in the spongy mesophyll.

The cells of the embryo contain aleurone grains and fixed oil.

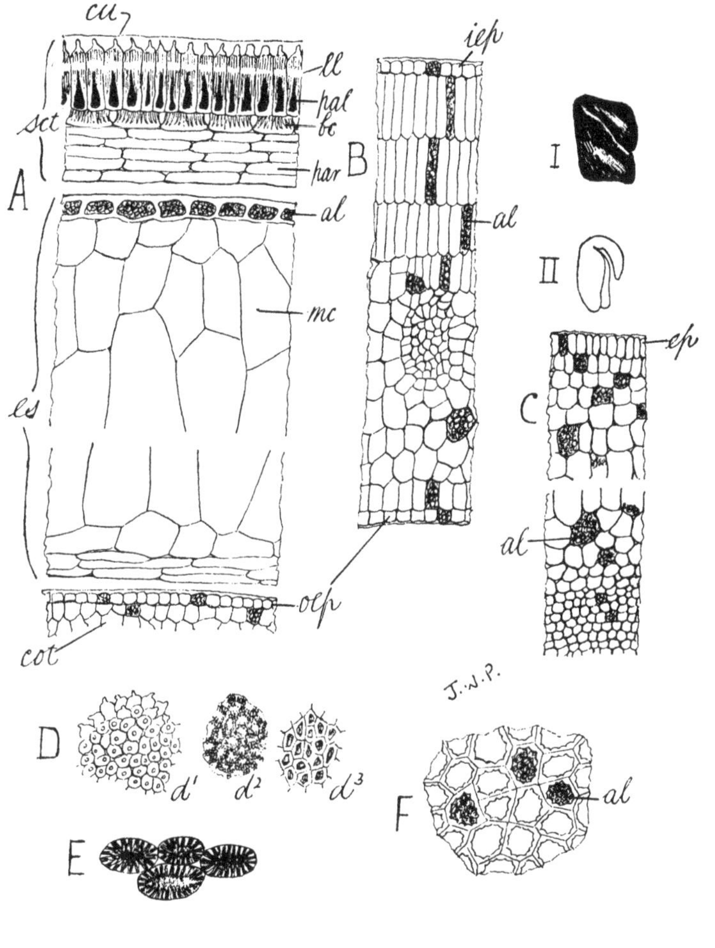

Fig. 15. FENUGREEK. I—Seed. II—Embryo. (A) Cross section through seed coat, endosperm, and part of cotyledon. (B) Cross section of cotyledon. (C) Cross section of radicle from epidermis to central zone. (D) Surface view of palisade cells of seed coat (d^1); as seen when microscope focused partly down (d^2); as seen when focused to base of cells (d^3). (E) Surface view of bearer cells. (F) Surface view of aleurone layer of endosperm.

al, aleurone; bc, bearer cells; cot, cotyledon; cu, cuticle; ep, epidermis; es, endosperm; iep, inner epidermis; ll, light line; mc, mucilage cells; oep, outer epidermis; pal, palisade cells; par, parenchyma; sct, seed coat.

XVII. MUSTARD, WHITE

Sinapis alba L.
Family: *Cruciferae*

[COMMONLY CALLED YELLOW MUSTARD]

GENERAL

An herbaceous, annual plant. Lower leaves pinnatifid, upper leaves sublyrate; leaves alternate along bristly branches, irregularly toothed, petiolate, hairy on both sides. Flowers in racemes, small, yellow; petals 4, cruciform; stamens tetradynamous; pistil bicarpellate. Fruit a bristly silique, roundish, ribbed, swollen at the seeds, and with a long ensiform beak; pods spreading in the raceme.

MORPHOLOGY

Seed yellowish, nearly globular, from about 1.5 to 3 mm (average of 25 seeds, 2.1 mm), minutely pitted, hilum scar, and with ridge and 2 parellel grooves or furrows externally indicating the position of the radicle in relation to the conduplicate cotyledons which form the bulk of the seed. Seed coat thin, endosperm meagre and imperceptible to the naked eye; large embryo, yellowish, with curved hypocotyl-radicle partially surrounded by 2 folded cotyledons. Seed becomes coated with mucilage when soaked in water.

Yellow mustard seeds are inodorous, and no odor results from crushing them in water (difference from black mustard seeds); taste, pungent.

HISTOLOGY

The tissues should be expanded with the aid of bleaching solution.

Seed

Cross section.

Seed Coat. This consists of: (a) thin cuticle; (b) epidermis consisting of large, mucilaginous cells nearly rectangular in shape, with convex outer walls; the mucilage is clearly stratified about a central column or axial cavity; (c) 2 layers of collenchyma cells; (d) a layer of palisade cells with pale straw-colored thickened walls in the lower part of the cells, and thin walls in the upper; and (e) 2 layers of collapsed parenchyma cells.

In the region of the hilum the palisade cells end. The parenchyma tissue underlying the palisade cells merges with that underlying the epidermis, and the whole tissue becomes several cells deep; in the outer part of the latter tissue small spiral vessels occur. (The structure is similar in black mustard seeds.) The collenchyma tissue increases from 2 layers to 3 or 4 layers toward the hilum region. There are no collenchyma cells in the immediate hilum region.

In surface view, the epidermal cells are polygonal in shape with thin, beaded walls; mucilage has central cavity. With polarized light each cell displays a striking, brilliant maltese cross effect. The thickening of the collenchyma cells is characteristic and noteworthy. The palisade cells are small, polygonal, and thick-walled; the thin-walled parts of the cells are not clearly distinguishable in surface view. The reticulations of the seed are only faintly discernible.

Endosperm. This consists of: (a) an outer layer of nearly quadrilateral to tangentially oblong, thick-walled cells containing aleurone grains; and (b) 6 or 7 layers of compressed cells.

In surface view, the aleurone cells are polygonal in shape and thick-walled.

Embryo. The embryo is large and consists of radicle and 2 folded cotyledons. The radicle consists of epidermis, several layers of isodiametric parenchyma cells, and a central core or zone of very small, thin-walled cells. The cotyledons consist of upper and lower epidermis, palisade cells, rudimentary spongy mesophyll, and rudimentary vascular tissue.

In surface view, daughter cells, usually in pairs, are a noteworthy feature of the epidermis of the cotyledons.

The cells of the embryo contain aleurone grains and fixed oil.

Ground White Mustard

Color light-yellow to pale brownish-yellow. The important identifying tissues are (a) fragments of the seed coat epidermis with mucilaginous cells; (b) collenchyma cells; (c) palisade cells with pale-colored thickening laid down on the inner and lower portions of their walls; (d) fragments of endosperm aleurone layer; and (e) fragments of the embryo with aleurone grains and fixed oil.

Ground Mustard

A blend of white and black mustard. See following chapter.

XVIII. MUSTARD, BLACK

Brassica nigra (L.) Koch
Family: *Cruciferae*

[COMMONLY CALLED BROWN MUSTARD]

GENERAL

An erect, herbaceous annual. Leaves alternate, petiolate, and variously shaped. Lower leaves large, rough, irregularly sinuate-dentate, lyrate-pinnate with the terminal lobe large and lower lobes small; rather hispid. Upper leaves smooth and moderately lobed. Uppermost leaves entire and narrow. Flowers in racemes, small, yellow; petals 4, cruciform; stamens tetradynamous; pistil bicarpellate. Fruit a silique, smooth, quadrangular, with a short, slender beak; pods nearly parallel to the axis of the raceme.

MORPHOLOGY

Seed dark reddish-brown, nearly globular, from about 1 to 1.6 mm (average of 25 seeds, 1.48 mm); finely but distinctly pitted, fine reticulations under microscope; hilum scar. Seed coat thin, endosperm imperceptible to the naked eye, embryo smaller but otherwise similar to that of white mustard. When soaked in water the seeds become coated with mucilage.

Black mustard seed is inodorous, but when crushed in water yields a sharp, piercing, irritating odor (presence of allyl isothiocyanate); taste, pungent.

HISTOLOGY

Seed

Cross section.

Seed Coat. This consists of: (a) thin cuticle; (b) an epidermal layer of mucilaginous cells tangentially oblong-rectangular in shape with convex outer walls; mucilage stratified, central column invisible or very faint; (c) a layer of large, thin-walled parenchyma cells; (d) a layer of palisade cells with dark-colored, thickened walls in the lower part, and thin walls in the upper; and (e) 1 or 2 layers of tangentially elongated pigment cells.

In surface view, the epidermal cells are polygonal in shape with thin, beaded walls; mucilage with central cavity. With polarized light each cell displays a bright maltese cross effect. The large parenchyma cells agree in size and and position with the reticulations of the seed coat. The reticulations of the seed coat are prominent and vary in size from about 88 to 133 microns, long axis. The palisade cells are small, polygonal, and thick-walled.

Endosperm. Consisting of (a) an outer layer of tangentially oblong, thick-walled cells containing aleurone grains; and (b) 3 or 4 layers of compressed cells.

In surface view, the aleurone cells are polygonal in shape and thick-walled.

Embryo. The embryo is similar to that described under white mustard.

Ground Black Mustard

Color, yellowish-brown. The important identifying tissues are (a) fragments of the seed coat epidermis with mucilaginous cells; (b) large parenchyma cells; (c) palisade cells with dark-colored thickening laid down on the inner and lower portions of their walls; (d) pigment cells; (e) fragments of endosperm aleurone layer; and (f) fragments of the embryo with aleurone grains and fixed oil.

Ground Mustard

To obtain the right taste and flavor it is the practice to blend white and black mustard, therefore, samples of pure mustard powder will contain the tissues of both white and black mustard seeds.

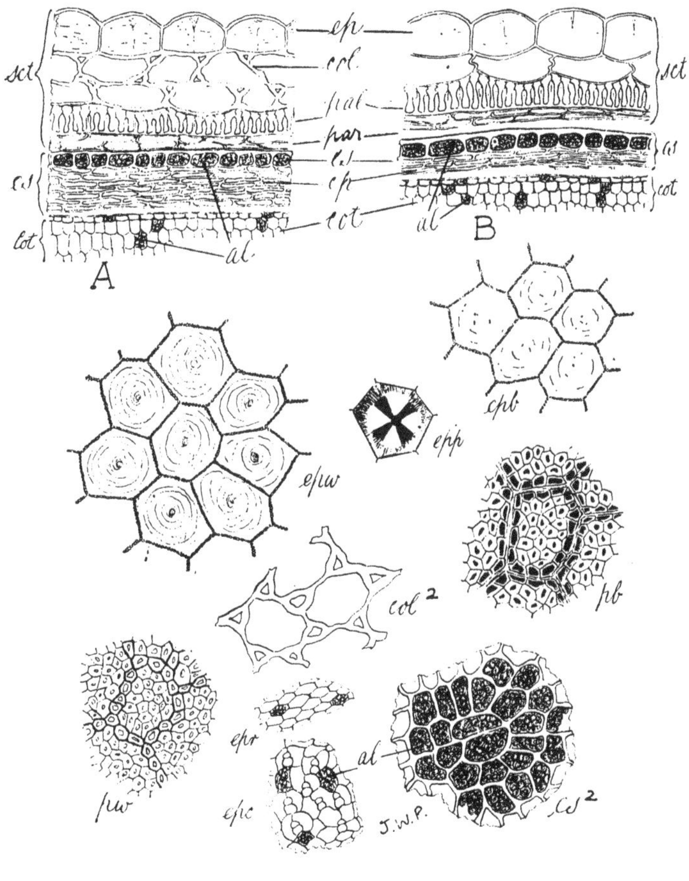

Fig. 16. MUSTARD SEED. (A) White. (B) Black Mustard; cross section through seed coat, endosperm, and part of cotyledon.

al, aleurone; col, collenchyma; cot, cotyledon; cp, compressed cells; ep, epidermis; es, endosperm; pal, palisade cells; par, parenchyma; sct, seed coat. Surface view: col², collenchyma; epb, epidermis of seed coat (black); epc, epidermis of cotyledon (white and black); epp, epidermis of seed coat with polarized light (white and black); epw, epidermis of seed coat (white); epr, epidermis of radicle (white and black); es², endosperm (white and black); pb, palisade cells, and prominent reticulation (black); pw, palisade cells, and faint reticulation (white).

XIX. POPPY

Papaver somniferum L.
Family: *Papaveraceae*

GENERAL

An annual herb reaching up to 4 feet in height, smooth, glaucous, with a branched stem. Leaves large, alternate, dull green, ovate, sinuate to dentate-serrate, clasping at the base. Flowers large, showy, varying in color from light pink to purple, and consisting of a receptacle borne on a long peduncle, erect, 2 green sepals, 4 colored petals, numerous stamens, syncarpous pistil, globose ovary bearing on its surface a radiate, sessile stigma. Fruit a capsule with numerous small seeds. Opium is obtained from this plant but the seeds are not narcotic.

The blue poppy seed of commerce may be the genuine blue poppy seed or it may be the white poppy seed artificially colored. The dye used for this purpose is water soluble and may be detected by shaking a few seeds in a little water contained in a test tube.

MORPHOLOGY

Seed grayish blue to dark blue, reniform, very small, about 1 to 1.25 mm in length, hilum distinct in a depression near one end; surface reticulated, the reticulations forming a raised-ridge network clearly visible with the aid of a hand lens. Seed coat thin; endosperm bulky; embryo consisting of radicle and 2 cotyledons embedded in the endosperm. Exudation of oil when pressure is applied to the seed with the finger nail.

Odor slight, pleasant, nutlike; taste agreeable and decidedly nutty.

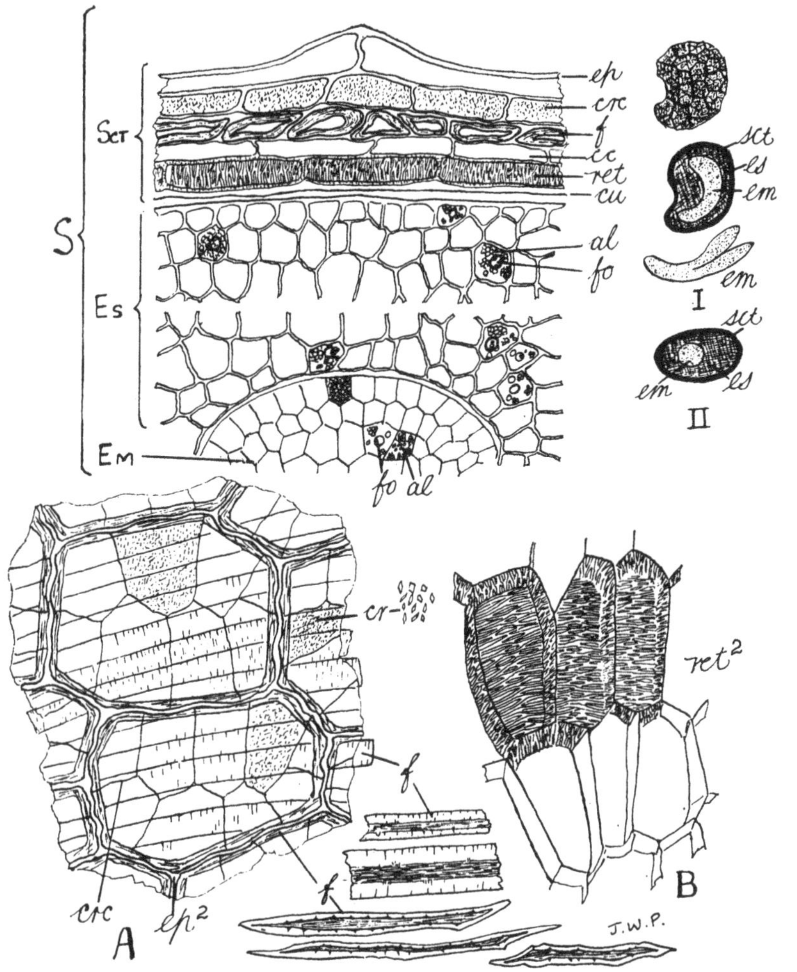

Fig. 17. POPPY SEED. I—Seed, whole; longitudinal section; and embryo. II—Cross section of seed. (S) Cross section of seed. (A) Surface view of seed coat. (B) Surface view of reticulated cells.

al, aleurone; cc, cross cells; cr, crystals; crc, crystal cells; cu, cuticle; em, embryo; ep, epidermis (ep^2, *surface view*)*; es, endosperm; f, fibers; fo, fixed oil; ret, reticulated cells* (ret^2, *surface view*)*; sct, seed coat.*

HISTOLOGY

Poppy seed is very small and sectioned with difficulty, but with care and patience sections which will permit a full study of the tissues can be obtained. Polarized light is very useful in an examination of the histological structure of this seed. The tissues can be expanded with the aid of bleaching solution.

Seed Coat

Cross section. Consisting of (a) an epidermis of tangentially-elongated, narrow, thin-walled cells; (b) a layer of parenchymatous cells containing numerous very small prismatic crystals of calcium oxalate; (c) a layer of thick-walled fibers, pronounced at the ridges of the seed coat. Fibers stain a weak blue with chlor-zinc-iodine following bleaching which would indicate that they are largely cellulose in composition; separation of the fibers is a difficult task; (d) a layer of parenchymatous cells; (e) a layer of reticulated cells. The reticulated cells are cellulose and stain blue with chlor-zinc-iodine; staining also permits a study of the detail of these cells; and (f) an inner cuticle. This cuticle is visible in an unstained section but its presence can be confirmed by staining a bleached section with chlor-zinc-iodine; the cuticle will be seen as a yellow membrane situated between the reticulated cells of the seed coat and the endosperm.

In surface view, the cells of the epidermis agree in size and shape with the deep surface reticulations of the seed. The ribs forming the reticulations consist of large, thick-walled fibers; the epidermal cells are shallow and thin-walled. The cells of the crystal layer are polygonal and contain numerous prismatic crystals of calcium oxalate; polarized light aids the study of these crystals. The fibers are broad, long, and lie at right angles to the underlying parenchyma and reticulated cells. The reticulated cells are polygonal in shape, and vary in size up to about 140 microns, long axis.

Endosperm and Embryo

The cells of the endosperm and embryo are polygonal and contain aleurone grains and globules of fixed oil. The embryo consists of radicle and 2 cotyledons.

XX. SESAME

Sesamum indicum L.
Family: *Pedaliaceae*

GENERAL

An annual, herbaceous plant with branched stem, 2 to 4½ feet or more in height, somewhat pubescent. Lower leaves opposite, often 3-lobed or parted; upper leaves alternate, oblong-lanceolate, serrate. Flowers about 1 inch long, pale rose or white, solitary, axillary, tubular-campanulate, 5-lobed. Fruit an oblong obtusely 4-sided, 2-loculed capsule (pod), containing small, flattish, pyriform seeds.

MORPHOLOGY

Unhulled Sesame Seed

Color varies with variety from yellowish-white to black through shades of gray and brown; flattened, pyriform; 2.5 to 3.5 mm in length, 1.5 to 2 mm in breadth, not more than 1 mm in thickness; raphe marked by a central longitudinal ridge on one side; a ridge around each side near the edge, distinct on some seed, indistinct on others; endosperm meagre; embryo large with radicle and 2 bulky cotyledons.

Hulled Sesame Seed

Divested of seed coat; creamy or pearly-white; flattened, pyriform; very slightly smaller than the unhulled seed; smooth, lustrous; and consisting of endosperm and embryo.

The product of the spice trade is usually the hulled seed.

Sesame seed has a very faint nutty odor, and an agreeable nutty taste.

HISTOLOGY

Seed Coat

Cross section. Tissues should be expanded with the aid of bleaching solution. The seed coat consists of (a) an epidermis of palisade or radially-elongated cells except at the ridges where the cells change in length and arrangement to form an outwardly-projecting pinnate group. The cells vary in size from about 30 to 60 microns on the sides of the seed, and from about 45 to 130 microns at the ridges; the cell walls are thin with the outer walls cutinized. Excepting at the ridges, each epidermal cell contains at its apex, within a very thin membranous tissue, a rosette-like group of calcium oxalate crystals. These crystal groups are brilliant with polarized light; (b) parenchymatous tissue with cells containing numerous prismatic crystals of calcium oxalate; at the ridges this tissue is fragmentary; and (c) an inner cuticle (stains yellow with chlor-zinc-iodine after bleaching).

Endosperm

Consisting of 3 to 4 layers of polygonal cells containing aleurone grains and globules of fixed oil.

Embryo

The embryo consists of radicle and 2 bulky cotyledons. The cotyledons are well developed and consist, in cross section, of an upper and lower epidermis of small rectangular cells, a palisade layer, polygonal parenchyma, and rudimentary vascular tissue. The cells contain aleurone grains and globules of fixed oil.

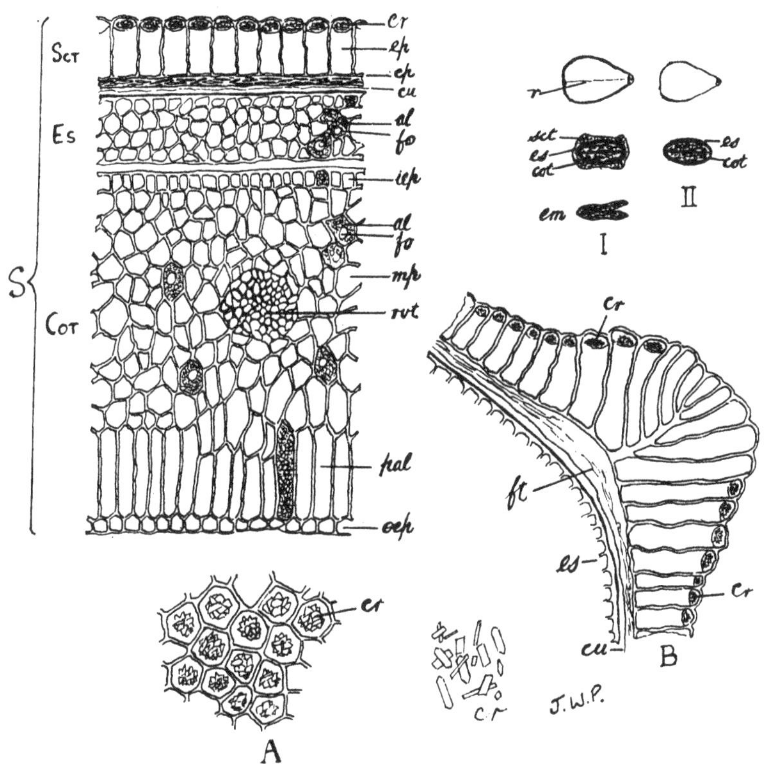

Fig. 18. SESAME. I—Unhulled seed. II—Hulled seed. (S) Cross section of seed. (A) Epidermis of seed coat with crystals. (B) Cross section of ridge of seed coat showing pinnate arrangement of epidermal cells.

al, aleurone; cot, cotyledon; cp, compressed cells; cr, crystals; cu, cuticle; em, embryo; ep, epidermis; es, endosperm; fo, fixed oil; ft, fragmentary tissue; iep, inner epidermal cells of cotyledon; mp, mesophyll of cotyledon; oep, outer epidermal cells; pal, palisade cells; r, raphe; rvt, rudimentary vascular tissue; sct, seed coat.

Umbelliferae

GENERAL

These are mostly perennial herbs. The name of this family is derived from the fact that the inflorescence of the plant is characteristically of the umbel type. The inflorescence is usually a compound umbel. Flowers small, regular, complete, and epigynous; calyx very small, inconspicuous, and seen only as 5 small teeth at the apex of the fruit. The ovary is inferior. Fruit a schizocarp, the ovary consisting of two fused carpels which separate at maturity into two 1-seeded mericarps which remain suspended for a time from a carpophore. The carpophore is divided in some species, partially divided in others, undivided in still others; it is a modified extension of the pedicel. Mericarps usually with prominent longitudinal ridges on the dorsal side, and tubes or vittae containing aromatic oil traversing the mesocarp.

Umbelliferae fruits employed as spices, and their main identifying characteristics, follow.

Anise: schizocarps ovoid; small; short hairs; numerous vittae; carpophore divided for a little more than half its length.

Caraway: dark colored; ridges prominent; 6 vittae; carpophore divides at the middle.

Celery: very small; smallest of the Umbelliferae spice fruits; schizocarps nearly circular; usually 6 vittae; carpophore divides near the top.

Coriander: schizocarps nearly globular with 10 straight and 10 wavy ridges; usually 2 vittae, commissural; carpophore undivided.

Cumin: hairy ridges, particularly the secondary ridges; 6 vittae; strong, heavy odor.

Dill: broadly oval or elliptical; conspicuous for wing-like lateral ridges; 6 vittae; carpophore divided.

Fennel: light-colored; largest of the Umbelliferae spice fruits; ridges very prominent; usually 6 vittae; carpophore divided.

Note: The number of vittae just given is the number to each mericarp. They are seen in a cross section of the larger fruits with the aid of a good hand lens, in the smaller fruits only with the aid of a microscope.

XXI. ANISE

Pimpinella anisum L.
Family: *Umbelliferae*

GENERAL

An annual herbaceous plant with an erect, smooth, branching stem rising to a height of 1 foot or more. Leaves petiolate, and varying from broadly triangular, cleft, and incised-serrate at the base of the plant to narrow segments, and trifid in the upper plant. Flowers small, white, and borne in large, compound umbels. Fruits are schizocarps with numerous vittae or oil tubes. Each mericarp contains a single seed.

MORPHOLOGY

Anise fruit, commonly called aniseed, is a schizocarp, ovoid, grayish-green to grayish-brown, 2 to 5 mm, short stylopodium, usually with pedicel, and consists of 2 mericarps attached to a carpophore divided a little over half its length.

The mericarp is broadly ovoid, flat or concave on the commissural side, convex on the dorsal side with 5 light-colored longitudinal ridges and numerous short hairs; the hairs are visible with the aid of a lens.

The pericarp contains many vittae or oil canals and encloses a single seed with a small embryo embedded in the endosperm toward the apex of the mericarp.

Anise has a characteristic, agreeable, aromatic odor, and a pleasant, aromatic taste.

HISTOLOGY

Pericarp

Cross section.

Exocarp. Consisting of a thin cuticle, a layer of epidermal cells oblong-rectangular in shape or nearly so, and numerous unicellular and some dicellular non-glandular, strongly papillose hairs varying in length to about 150 microns, blunt at the apex and broad at the base, mostly curved. Some of the dicellular hairs contain a brownish substance.

In surface view, the cuticle is striated; the epidermal cells are polygonal in shape, and vary in size to about 40 microns, long axis. Head on, the hairs appear as small circles ringed with dots. Stomata are present.

Mesocarp. Consisting of thin-walled, isodiametric or elliptical parenchyma cells in the outer mesocarp, and 3 layers of tangentially elongated cells in the inner mesocarp. Toward the apex of the mericarp the inner mesocarp cells adjoining the endocarp in the central commissural region are sclerenchymatous.

Endocarp. Consisting of a layer of tangentially-elongated, thin-walled cells. Toward the apex of the mericarp the endocarp cells in the central commissural region are nearly quadrilateral in shape and are sclerenchymatous.

In surface view, with the exceptions given below, the endocarp cells are transversely long, narrow, and thin-walled. On the commissural side of the mericarp, toward the base, the endocarp cells are longitudinally elongated and thin-walled; in the central commissural region they vary in shape from nearly quadrilateral to transversely long and narrow, are mostly thin-walled but some are weakly sclerenchymatous; toward the apex they vary in shape from nearly quadrilateral to longitudinally oblong-rectangular and are strongly sclerenchymatous.

Fibrovascular Bundles. Five fibrovascular bundles traverse the outer mesocarp and form the ridges of the mericarp. Both spiral and reticulate vessels occur; the xylem fibers are pitted and vary in length. The phloem fibers are long and narrow.

Vittae. There are many dorsal vittae, and 2 commissural vittae. The dorsal vittae are small, up to about 165 microns, tangential axis; the commissural vittae are large, up to about 495 microns, tangential axis. The vittae are yellowish-brown, elliptical, schizogenous, secretory cavities.

In surface view, the vittae are yellowish-brown, tubular tissues consisting of small, polygonal, cutinized cells; membranous cross walls occur at intervals.

Seed

Cross section.

Seed Coat. This consists of a single layer of epidermal cells on the dorsal side, an epidermal layer and several underlying layers of parenchyma cells on the commissural side; the raphe is centrally situated in the latter tissue. The epidermal cells are tangentially oblong-rectangular; the parenchyma cells are tangentially elliptical except around the raphe where they are small and isodiametric or irregular in shape. The outer walls of the epidermal cells are cutinized. In surface view, the epidermal cells are polygonal in shape. An inner, thin cuticle is present.

Endosperm. This consists of polygonal cells with moderately thick, colorless walls. The cell contents consist of aleurone grains and globules of fixed oil. There is a rosette crystal of calcium oxalate embedded in many of the larger aleurone grains.

Embryo. The embryo varies in length from about 1350 to 1800 microns, and consists of radicle and 2 cotyledons; the radicle lies toward the apex of the seed. The cells of the embryo contain aleurone grains and globules of fixed oil.

Pedicel

The pedicel consists of a smooth cuticle, an epidermis with occasional hairs, cortex, vittae, and a cylindrical core of fibrovascular tissue. In cross section, the epidermal cells are oblong-rectangular in shape, or nearly so, with thick outer walls. The cortex cells are small, isodiametric, and thin-walled. A few weakly developed collenchyma cells underlie the epidermis near the vittae. The vittae are 2 in number. In surface view, the epidermal cells are polygonal in shape excepting over collenchyma tissue where they are oblong-rectangular. Stomata are fairly numerous.

Carpophore

The carpophore is a modified extension of the pedicel and divides for just over half its length.

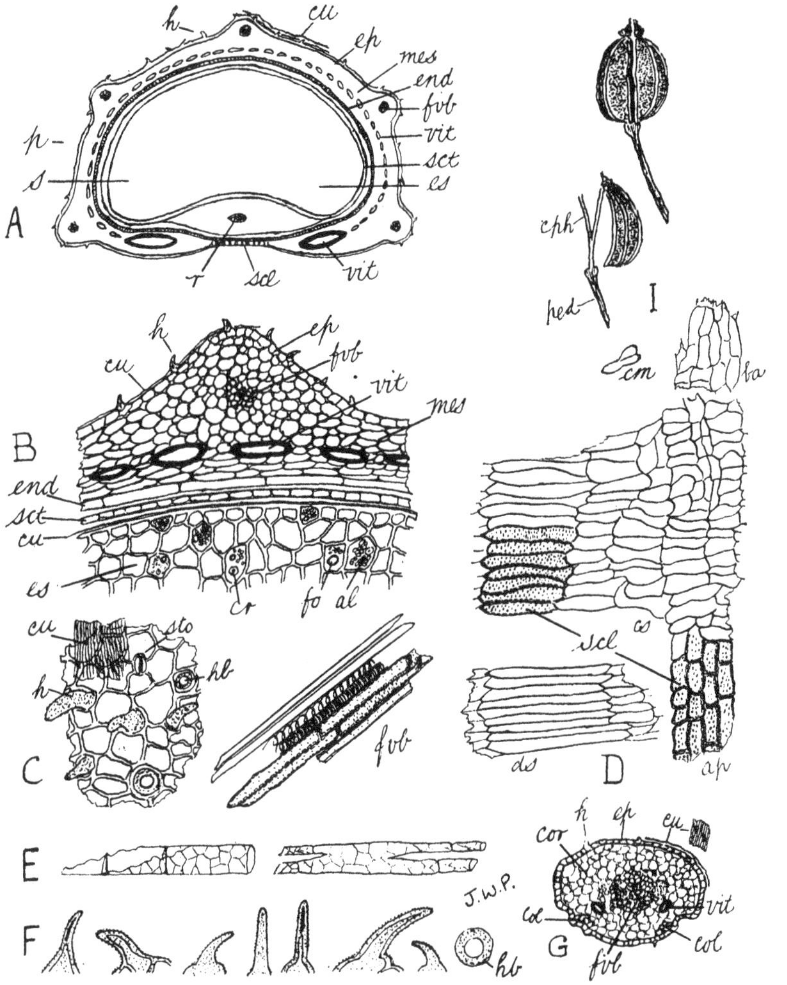

Fig. 19. ANISE. I—Schizocarp (above); mericarp (below). (A) Cross section of mericarp. (B) Cross section through ridge. (C) Epidermis of pericarp, surface view. (D) Endocarp, surface view. (E) Longitudinal view of vitta, simple and branched. (F) Hairs. (G) Cross section of pedicel.

al, aleurone; ap, apex; ba, base; col, collenchyma; cor, cortex; cph, carpophore; cr, crystal; cs, commmissural side; cu, cuticle; ds, dorsal side; em, embryo; end, endocarp; ep, epidermis; es, endosperm; fo, fixed oil; fvb, fibrovascular bundle; h, hair; hb, hair base; mes, mesocarp; p, pericarp; ped, pedicel; r, raphe; s, seed; scl, sclerenchyma; sct, seed coat; sto, stoma; vit, vitta.

XXII. CARAWAY

Carum carvi L.
Family: *Umbelliferae*

GENERAL

This is a biennial herbaceous plant reaching a height of about 2 feet. Leaves compound with linear segments. Flowers small, white, and produced in compound umbels. Fruits schizocarps; each mericarp contains a single seed.

MORPHOLOGY

The dried product of commerce consists almost entirely of separated mericarps. Mericarps hard, sharp, light to dark brown, 4 to 6 mm in length, some straight but mostly curved, short stylopodium, flat on the commissural side, convex on the dorsal side with 5 prominent, light-colored, longitudinal ridges or ribs; pedicel rarely present.

The pericarp contains 4 dorsal and 2 commissural vittae, and encloses a single seed with a small embryo embedded in the endosperm toward the apex.

The carpophore divides at the middle.

Caraway has a characteristic, agreeable odor, and an aromatic, warm, somewhat sharp taste.

HISTOLOGY

Pericarp.

Cross section.

Excarp. Consisting of cuticle, and a layer of epidermal cells tangentially oblong-rectangular in shape or nearly so; the epidermal cells over the ridges have thick outer walls.

In surface view, the epidermal cells are somewhat wavy-walled and polygonal in shape except over the ridges where they are straight-walled and longitudinally oblong or nearly so. The cell walls are beaded. Prismatic crystals of calcium oxalate occur in most of the cells. A few stomata are present.

Mesocarp. Consisting of tangentially elliptical cells in the outer mesocarp, and 3 or 4 layers of closely compressed, yellowish cells in the inner mesocarp.

Endocarp. Consisting of a layer of tangentially oblong, broad cells except in the central commissural region of the mericarp where they become radially oblong or quadrilateral in shape.

In surface view, the endocarp cells are transversely elongated except in the central commissural region of the mericarp where they are longitudinally oblong, and at the apex where they become quadrilateral or polygonal in shape. On the dorsal side of the mericarp the endocarp cells vary in size from about 45 to 135 microns, long axis, and from about 14 to 22 microns, short axis; on the commissural side they vary in size from about 45 to 180 microns, long axis, and from about 9 to 22 microns, short axis. The cell walls are thin except in the central and apex commissural regions where they become, for the most part, sclerenchymatous.

Fibrovascular Bundles. Five fibrovascular bundles traverse the mesocarp; 3 of these bundles form the dorsal ridges, and 2 the commissural or marginal ridges. In the basal and apex regions of the mericarp a number of pitted sclerenchyma cells of various shapes and sizes accompany the fibrovascular bundles of the commissural ridges.

Vittae. There are 4 dorsal, and 2 commissural, vittae. The dorsal vittae vary from about 220 to 420 microns, long axis; the commissural vittae from about 198 to 288 microns, long axis. The vittae are yellowish-brown, elliptical, schizogenous, secretory cavities; in surface view, they are yellowish-brown, tubular tissues consisting of cutinized, polygonal cells; membranous cross walls occur at intervals.

Unlike other fruits of the Umbelliferae considered in this work, there are, in addition to the vittae discussed above, 5 small, schizogenous, secretory cavities in the outer mesocarp, 1 at each ridge above and adjoining the fibro vascular bundle. These vittae are isodiametric or elliptical in shape, and vary in size from about 20 to 45 microns diameter or long axis. In surface view, they are yellowish-brown, narrow, tubular tissues traversing the length of the mericarp.

Seed

Cross section.

Seed Coat. Consisting of a layer of epidermal cells on the dorsal side of the seed, and a layer of epidermal cells with underlying parenchyma on the commissural side; the raphe is centrally situated in the latter tissue. The epidermal cells are tangentially oblong-rectangular in shape, and have cutinized outer walls; the parenchyma cells are thin-walled, small, and compressed. In surface view, the epidermal cells are polygonal in shape. An inner, thin cuticle is present.

Endosperm. Consisting of polygonal cells with moderately thick, colorless walls. The cell contents consist of aleurone grains and globules of fixed oil. There is a rosette crystal of calcium oxalate embedded in many of the larger aleurone grains.

Embryo. The embryo varies in length from about 1800 to 2000 microns and consists of radicle and 2 cotyledons. It is embedded in the endosperm with the radicle toward the apex of the seed. The cells contain aleurone grains and globules of fixed oil.

Pedicel

A pedicel is present on very few caraway fruits. It is similar to that described under fennel.

Carpophore

The carpophore is a modified extension of the pedicel and is divided for about half its length.

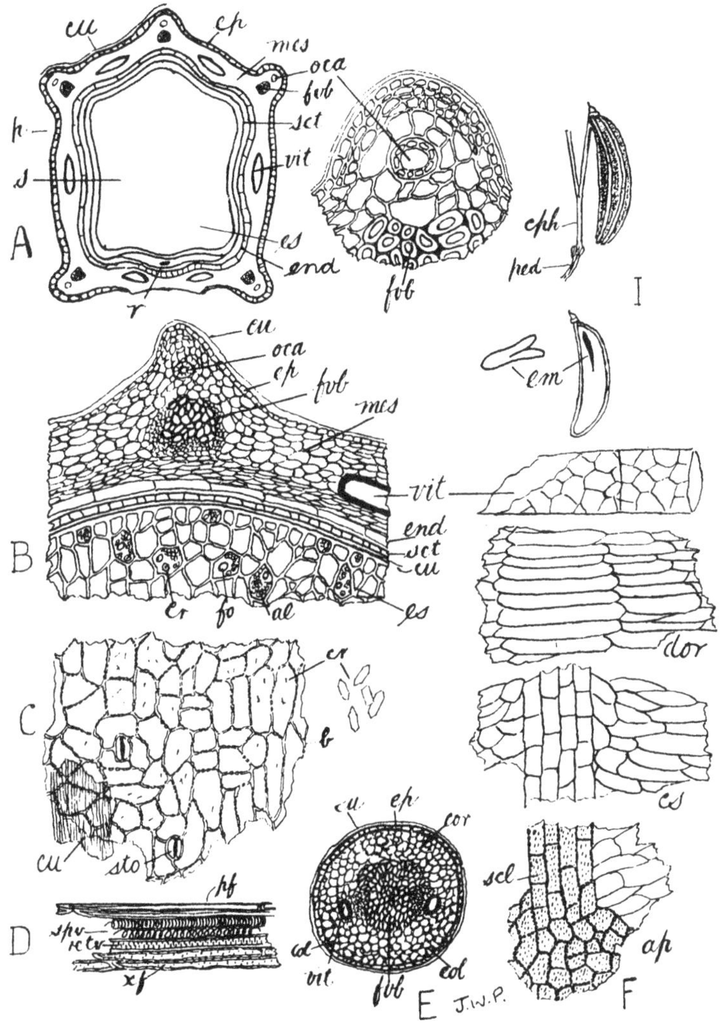

Fig. 20. CARAWAY. I—Mericarp. (A) Cross section of mericarp. (B) Cross section through ridge. (C) Epidermis of pericarp, surface view. (D) Fibro vascular tissue. (E) Cross section of pedicel. (F) Endocarp, surface view.

al, aleurone; ap, apex; b, beaded walls; col, collenchyma; cor, cortex; cph, carpophore; cr, crystals; cs, commissural side; cu, cuticle; dor, dorsal; em, embryo; end, endocarp; ep, epidermis; es, endosperm; fo, fixed oil; fvb, fibrovascular bundle; mes, mesocarp; oca, oil canal; p, pericarp; ped, pedicel; pf, phloem fibers; r, raphe; retv, recticulated vessel; s, seed; scl, sclerenchyma; sct, seed coat; spv, spiral vessel; sto, stoma; vit, vitta; xf, xylem fibers.

XXIII. CELERY

Apium graveolens L.
Family: *Umbelliferae*

GENERAL

Cultivated celery is a biennial herb reaching a height of about 18 inches. Root leaves well developed, succulent, numerous; leaflets usually 3 pair and a terminal one, ternately compound, coarsely toothed or cleft. Flowers white, very small, and produced in compound umbels. Fruits schizocarps; each mericarp contains a single seed.

MORPHOLOGY

The dried celery fruits of the spice trade, commonly but erroneously called celery seeds, consist of schizocarps and separated mericarps, many with pedicel attached.

The schizocarp is very small, only a very occasional fruit exceeding 1.6 mm, nearly circular, grayish-brown, and consists of 2 mericarps attached to a carpophore divided near the top.

The mericarp is flat to slightly concave on the commissural side, and convex on the dorsal side with 5 light-colored longitudinal ridges.

The pericarp usually contains 4 dorsal and 2 commissural vittae but occasionally more, and encloses a single seed with a small embryo embedded in the endosperm toward the apex of the mericarp.

The fruit has the well-known celery aroma and taste, the latter warm and slightly bitter.

HISTOLOGY

Pericarp

Cross section.

Exocarp. This consists of a cuticle and a layer of epidermal cells, many of which are papillose.

In surface view, the cuticle is strongly striated; the epidermal cells are irregular in shape with thin, wavy walls except over the ridges of the mericarp where they become longitudinally elongated with straighter, thicker walls. Head on, the papillae appear as small circles with wrinkled rims. Stomata are present but not abundant.

Mesocarp. This consists of tangentially elliptical parenchyma cells in the outer mesocarp, and a layer of tangentially elongated cells with pale-yellowish walls in the inner mesocarp; some of the latter cells become sclerenchymatous in the central and apex commissural regions of the mericarp.

Endocarp. This consists of a layer of tangentially elongated, narrow, pale-yellowish cells.

In surface view, the endocarp cells are elongated, narrow, and thin-walled; mostly transverse, occasionally parqueted in the basal and apex regions of the mericarp, longitudinally oblong in the central commissural region.

Fibrovascular Bundles. Five fibrovascular bundles traverse the outer mesocarp; 3 of these form the dorsal ridges, and 2 the commissural ridges of the mericarp. Lignified, reticulated cells accompany the bundles; they are particularly numerous in the apex region. The xylem vessels are spiral.

Vittae. Usually there are 4 dorsal, and 2 commissural, vittae, but additional vittae are occasionally present. The vittae are yellowish-brown, elliptical, schizogenous, secretory cavities varying from about 117 to 166 microns, long axis.

In surface view, the vittae are yellowish-brown, tubular tissues consisting of cutinized, polygonal cells; membranous cross walls occur at intervals.

Seed

Cross section.

Seed Coat. This consists of a layer of epidermal cells on the dorsal side of the seed, and a layer of epidermal cells with underlying parenchyma on the commissural side; the raphe is centrally situated in the latter tissue. The epidermal cells are tangentially oblong-rectangular in shape and have cutinized outer walls; the parenchyma cells are thin-walled and elliptical in shape except around the raphe where they are isodiametric or irregular.

In surface view, the epidermal cells are polygonal in shape. An inner, thin, smooth cuticle is present.

Endosperm. This consists of polygonal cells with moderately thick, colorless walls. The cell contents consist of aleurone grains and globules of fixed oil. There is a rosette crystal of calcium oxalate embedded in many of the larger aleurone grains.

Embryo. The embryo varies in length from about 400 to 500 microns, and consists of radicle and 2 cotyledons. It is embedded in the endosperm with the radicle toward the apex of the seed. The cells contain aleurone grains and fixed oil.

Pedicel

Except for the absence of palisade-like cells in the cortex, the presence of only a small amount of collenchyma tissue, and a striated cuticle, the pedicel is similar to that described under fennel.

Carpophore

The carpophore is a modified extension of the pedicel and is divided from a point very close to the apex of the mericarp.

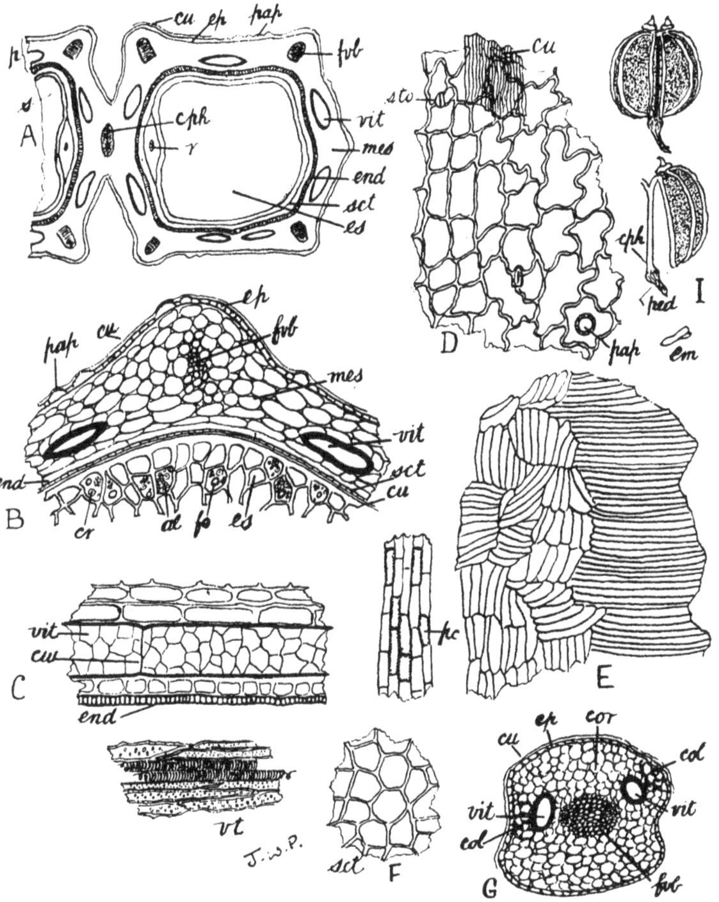

Fig. 21. CELERY. I—Schizocarp (above); mericarp (below). (A) Cross section of mericarp. (B) Cross section through ridge. (C) Longitudinal view of vitta. (D) Epidermis of pericarp. (E) Endocarp. (F) Seed coat. [D, E, and F, in surface view.] (G) Cross section of pedicel.

al, aleurone; col, collenchyma; cor, cortex; cph, carpophore; cr, crystal; cu, cuticle; cw, cross wall; em, embryo; end, endocarp; ep, epidermis; es, endosperm; fo, fixed oil; fvb, fibrovascular bundle; mes, mesocarp; p, pericarp; pap, papilla; ped, pedicel; pc, pitted cell; r, raphe; s, seed; sct, seed coat; sto, stoma; vit, vitta; vt, fibrovascular tissue.

XXIV. CORIANDER

Coriandrum sativum L.
Family: *Umbelliferae*

GENERAL

This is an annual herb growing to a height of about 2 feet. Lower leaves broad and deeply segmented or lobed; upper leaves composed of very narrow, linear segments. Flowers small, pinkish, and produced in compound umbels. Fruits schizocarps, spherical and ridged.

MORPHOLOGY

Dried coriander fruit is a nearly globular schizocarp, yellowish-brown, 3 to 5 mm in diameter, with 10 straight and 10 wavy longitudinal ridges, a depression at the base, 5 small calyx teeth and a conical stylopodium at the apex. The schizocarp of coriander consists of 2 mericarps attached to an undivided carpophore.

The mericarp is concave on the commissural side, and convex on the dorsal side.

The pericarp is without vittae on the dorsal side, and usually with 2 vittae on the commissural side but occasionally more. The pericarp encloses a single seed with a small embryo embedded in the endosperm toward the apex of the seed.

Coriander has a slight, fragrant odor, and a pleasantly aromatic taste.

HISTOLOGY

Pericarp

Cross section.

Exocarp. This consists of cuticle, an epidermis of thick-walled cells tangentially oblong-rectangular in shape or nearly so, and 2 supporting layers of thick-walled, compressed cells.

In surface view, the cuticle is striated; the epidermal cells are polygonal in shape, and vary in size up to about 45 microns, long axis, and have thick, pitted walls. Each cell contains 1 or more prismatic crystals, or a rosette-like group of microcrystals of calcium oxalate. Stomata are present.

Mesocarp. This includes an outer mesocarp consisting of (a) about 4 layers of tangentially oblong, thin-walled, parenchyma cells up to about 45 microns, long axis; (b) a layer of large, thin-walled, parenchyma cells which become palisade-like in the ridge regions; and (c) small, irregular or polygonal parenchyma cells adjoining a fiber layer.

There is also a fiber layer consisting of densely packed sclerenchyma fibers. Groups of fibers cross one another at various angles, frequently right angles. In surface view, the fibers follow a sinuous path.

Finally, there is an inner mesocarp consisting of (a) about 3 layers of thin-walled, tangentially elongated cells up to about 90 microns, long axis; and (b) 2 layers of thick-walled, pitted, tangentially elliptical cells up to about 90 microns, long axis; the walls of the cells adjoining the endocarp are the thickest of this tissue.

Endocarp. This consists of a layer of tangentially long, narrow cells, excepting where parqueting occurs.

In surface view, consisting of elongated, narrow cells up to about 300 microns, long axis, and 6 microns, short axis; transverse, occasionally parqueted.

Fibrovascular Bundles. Unlike the other Umbelliferae fruits considered in this work, the fibrovascular tissue is not conspicuous in cross section. It adjoins the fiber layer and is best studied in longitudinal section. Pitted, reticulated, and spiral vessels occur in the xylem.

Vittae. There are no dorsal vittae. Two commissural vittae are usually present, but, occasionally, additional commissural vittae occur. The vittae are elliptical, schizogenous, secretory cavities varying in size from about 150 to 300 microns, long axis, and from about 60 to 75 microns, short axis. They are flanked by thick-walled cells.

In surface view, the vittae are yellowish-brown, broad, tubular tissues consisting of small, polygonal cells with beaded walls; there are no cross walls.

Seed

Cross section.

Seed Coat. This consists of a layer of epidermal cells on the dorsal side, and a layer of epidermal cells with underlying parenchyma on the commissural side; the raphe is centrally situated in the latter tissue. The epidermal cells are thick-walled, and tangentially oblong-rectangular in shape excepting around the raphe where they are isodiametric or irregular. The outer walls of the epidermal cells are cutinized.

In surface view, the epidermal cells are polygonal in shape and thick-walled. An inner, thin, cuticle is present.

Endosperm. This consists of polygonal cells with moderately thick, colorless walls. The cell contents consist of aleurone grains and globules of fixed oil. A rosette crystal of calcium oxalate is embedded in many of the larger aleurone grains.

Embryo. The embryo varies in length from about 1000 to 1150 microns, and consists of radicle and 2 cotyledons. It is embedded in the endosperm with the radicle toward the apex of the seed. The cell contents are aleurone grains and globules of fixed oil.

Pedicel

The pedicel is similar to that described under fennel excepting that few, if any, palisade-like cells occur in the cortex parenchyma; the outer walls of the epidermal cells are thick, and the cuticle is faintly striated.

Carpophore

The carpophore is a modified extension of the pedicel and is not divided.

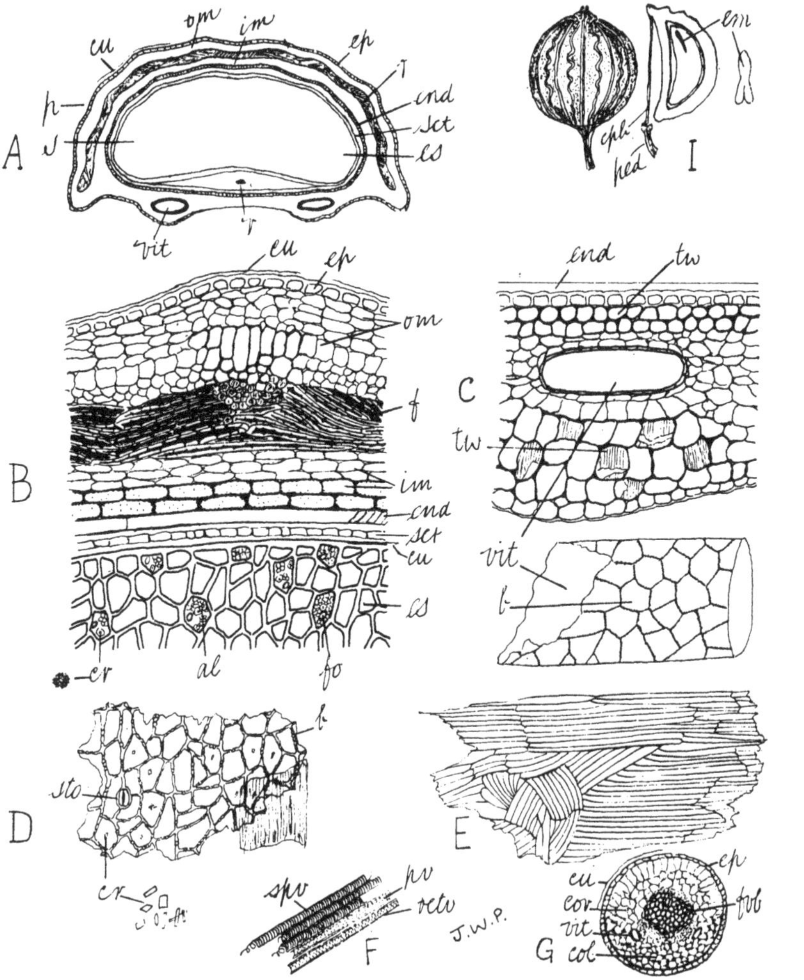

Fig. 22. CORIANDER. I—Schizocarp (left); mericarp (right). (A) Cross section of mericarp. (B) Cross section through primary ridge. (C) Cross section through commissural vitta. (D) Epidermis of pericarp, surface view. (E) Endocarp, surface view. (F) Vascular tissue. (G) Cross section of pedicel.

al, aleurone; b, beaded walls; col, collenchyma; cor, cortex; cph, carpophore; cr, crystals; cu, cuticle; em, embryo; end, endocarp; ep, epidermis; es, endosperm; f, fiber layer; fo, fixed oil; fvb, fibrovascular bundle; im, inner mesocarp; om, outer mesocarp; p, pericarp; ped, pedicel; pv, pitted vessel; r, raphe; retv, reticulated vessel; s, seed; sct, seed coat; spv, spiral vessel; sto, stoma; tw, thickened walls; vit, vitta.

XXV. CUMIN

Cuminum cyminum L.
Family: *Umbelliferae*

GENERAL

This is an annual, low-growing herb with finely divided leaves and small, white or rose-colored flowers produced in compound umbels. Fruits schizocarps; each mericarp contains a single seed.

MORPHOLOGY

Schizocarp ovoid-elongated, frequently with pedicel, stylopodium divided, yellowish-brown, 3 to 6.5 mm in length, and consisting of 2 mericarps attached to a carpophore divided from a point near the base.

Mericarp concave on the commissural side, convex on the dorsal side; 5 prominent, primary longitudinal ridges with a number of short hairs; 4 less distinct, secondary longitudinal ridges with numerous long hairs.

The pericarp contains 4 dorsal and 2 commissural vittae, and encloses a single seed with embryo embedded in the endosperm.

Cumin has a strong, heavy, distinctive odor, and a warm, aromatic taste.

HISTOLOGY

Pericarp

Cross section.

Exocarp. This consists of a thin cuticle and a layer of epidermal cells oblong-rectangular in shape or nearly so, with hairs occurring over the ridges. The hairs of the secondary ridges are much longer than those of the primary ridges, occasionally reaching 2 mm. The hairs are multicellular, multiseriate, broad at the base and rounded at the apex.

In surface view, the cuticle is striated. The epidermis of the primary ridges and the channels consists of irregular-shaped cells with thin, wavy or sinuous walls; stomata are present. The epidermis of the secondary ridges consists of cells longitudinally elongate-irregular in shape with moderately thick walls; there are no stomata.

Mesocarp. This consists of tangentially elliptical parenchyma cells with thin, colorless walls in the outer mesocarp, and tangentially elliptical cells with somewhat thicker, pale-yellowish walls in the inner mesocarp.

Endocarp. This consists of a layer of tangentially elongated, narrow cells with pale-yellowish walls. The cell walls are thin except in the apex region of the commissural side of the mericarp where they are thick and scerenchymatous.

In surface view, the thin-walled cells are transversely elongated and vary in size up to about 250 microns, long axis, and up to 14 microns, short axis. The sclerenchymatous cells vary in shape, size and arrangement; some are transversely elongated and narrow, others from transversely oblong to irregular and varying in width; those of the commissural apex region are longitudinally oblong.

Fibrovascular Bundles. Five fibrovascular bundles traverse the outer mesocarp and form the 5 primary ridges of the mericarp. Both spiral and reticulate vessels occur; the fibers are pitted.

Vittae. There are 6 vittae. Four of these give rise to the secondary ridges of the mericarp, 2 are commissural vittae. The vittae are yellowish-brown, schizogenous, secretory cavities. The dorsal vittae are up to about 250 microns, tangential axis, the commissural vittae to about 300 microns. Occasionally an additional vitta is present. When this occurs, the extra vitta usually adjoins one of the dorsal vittae, each being smaller than the customary dorsal vittae.

In surface view, the vittae are yellowish-brown, tubular tissues consisting of small, polygonal, cutinized cells; membranous cross walls occur at intervals.

Seed

Cross section.

Seed Coat. This consists of a layer of epidermal cells on the dorsal side, a layer of epidermal cells and underlying parenchyma on the commissural side; the raphe is centrally situated in the latter tissue. The epidermal cells are tangentially oblong-rectangular in shape with cutinized outer walls. The parenchyma cells are thin-walled and elliptical except around the raphe where they are isodiametric or irregular in shape.

In surface view, the epidermal cells of the seed coat are polygonal in shape; the outer cells walls are thicker than the inner, particularly so on the commissural side. Each cell contains a prismatic crystal of calcium oxalate. An inner, thin cuticle is present.

Endosperm. This consists of polygonal cells with moderately thick, colorless walls. The cells contain aleurone grains and globules of fixed oil. A small, rosette crystal of calcium oxalate is embedded in many of the larger aleurone grains.

Embryo. The embryo is about 1400 microns in length, and consists of radicle and 2 cotyledons. It is embedded in the endosperm with the radicle toward the apex of the seed. The cell contents are aleurone grains and globules of fixed oil.

Pedicel

The pedicel consists of cuticle, epidermis with stomata, cortex, 2 vittae, and fibrovascular tissue. There is no collenchyma tissue, but parenchyma cells with somewhat thickened walls fan out from the vittae to the epidermis. The remaining cortex tissue consists of 2 layers of thin palisade-like cells, and thinwalled, isodiametric or irregular cells. A large, cylindrical mass of fibrovascular tissue forms the bulk and core of the pedicel.

Carpophore

The carpophore is a modified extension of the pedicel and is divided from a point near the base of the mericarp.

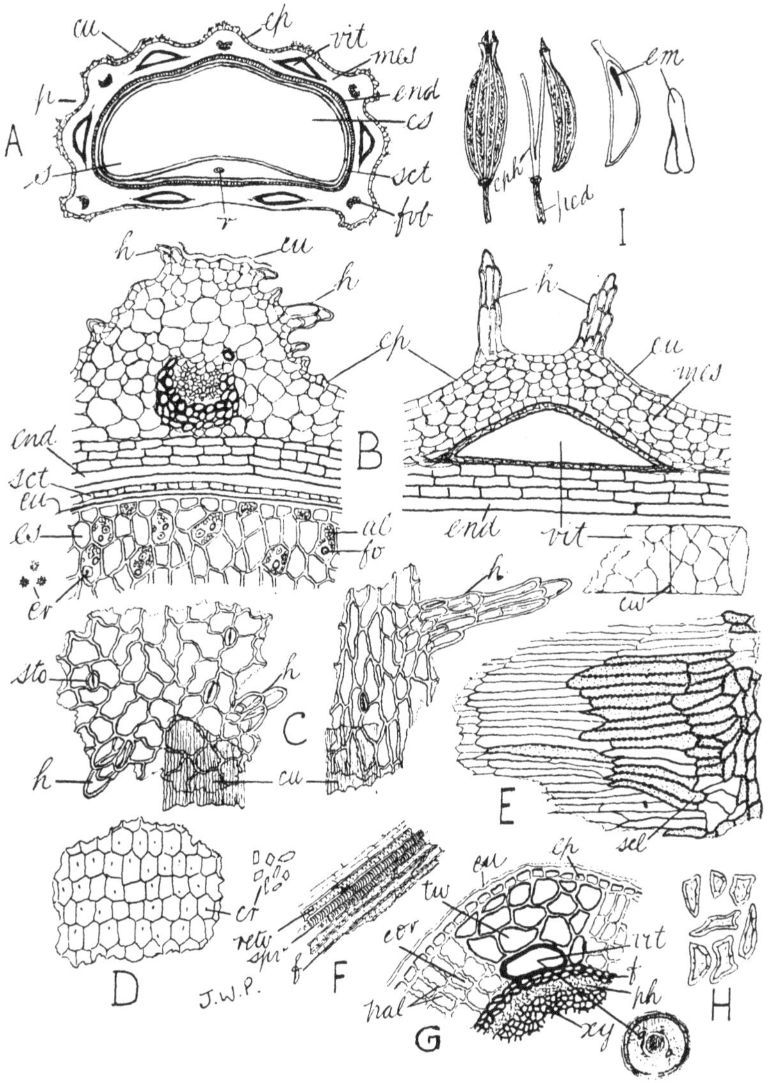

Fig. 23. CUMIN. I—Schizocarp (left); mericarp (right). (A) Cross section of mericarp. (B) Cross section through ridge (left); through dorsal vitta (right). (C) Surface view of epidermis of pericarp over primary ribs and channels (left); over secondary ribs (right). (D) Seed coat, surface view. (E) Endocarp, commissural side, apex region, surface view. (F) Fibrovascular tissue. (G) Cross section of pedicel. (H) Sclerenchyma cells occurring in dorsal mesocarp of apex region of pericarp.

al, aleurone; cor, cortex; cph, carpophore; cr, crystal (s); cu, cuticle; cw, cross wall; em, embryo; end, endocarp, ep, epidermis; es, endosperm; f, fibers; fo, fixed oil; fvb, fibrovascular bundle; h, hair; mes, mesocarp; p, pericarp; pal, palisade cells; ped, pedicel; ph, phloem; r, raphe; retv, reticulated vessel; s, seed; scl, sclerenchyma; sct, seed coat; spv, spiral vessel; sto, stoma; tw, thick-walled cells; vit, vitta; xy, xylem.

XXVI. DILL

Anethum graveolens L.
Family: *Umbelliferae*

GENERAL

This is an annual herb reaching a height of 3 or 4 feet, erect stem, finely divided light green leaves, and compound umbels of small yellow flowers. Fruits schizocarps; each mericarp contains a single seed. All parts of the plant are aromatic.

MORPHOLOGY

Schizocarp broadly oval or elliptical, light brown, short stylopodium, 3 to 5 mm long, 2 to 2.5 mm wide, and consisting of 2 mericarps attached to a divided carpophore. The commercial product consists mostly of mericarps with few pedicels present.

Mericarp concave on the commissural side, convex on the dorsal side, with 3 prominent, sharp, longitudinal dorsal ribs and 2 pale yellowish-brown, longitudinal, winglike lateral ribs.

The pericarp contains 4 dorsal vittae and 2 commissural vittae, and encloses a single seed with embryo embedded in the endosperm.

Dill has an aromatic odor faintly resembling that of caraway, and a warm, slightly sharp taste akin to caraway.

HISTOLOGY

Pericarp

Cross section.

Exocarp. This consists of cuticle and a layer of epidermal cells tangentially oblong-rectangular in shape or nearly so, with stomata.

In surface view, the cuticle is striated; the epidermal cells are polygonal.

Mesocarp. This consists of a layer of tangentially oblong-rectangular cells except some isodiametric and irregular parenchyma cells near the fibrovascular bundles, and several tangentially elliptical, reticulated sclerenchymatous cells flanking the fibrovascular bundles on the inner side toward the endocarp.

Endocarp. This consists of a layer of tangentially oblong-rectangular cells except (a) where parqueting occurs, and (b) in the central commissural region where the cells are nearly quadrilateral in shape. The cell walls are thin.

In surface view, the endocarp cells are elongated and narrow, mostly transverse, parqueted in the basal and apex regions of the mericarp, and longitudinally oblong in the central commissural region toward the apex.

Fibrovascular Bundles. Five fibrovascular bundles are present; 3 of these traverse the outer mesocarp and form the dorsal ridges or ribs, 2 are situated in the lateral ridges.

The lateral ridges are wide, often referred to as wings, and are characteristic of dill. The fibrovascular bundles occurring in the lateral ridges are large; they are flanked on the inner or endocarp side by reticulated, sclerenchymatous cells, and on the outer or edge side of the ridge by sclerenchyma cells wich thick, pitted walls.

Both spiral and reticulate vessels occur in the xylem of dill.

Vittae. There are 4 dorsal and 2 commissural vittae. The dorsal vittae vary to about 203 microns, tangential axis, the commissural vittae to about 250 microns. The vittae are yellowish-brown, schizogenous secretory cavities.

In surface view, the vittae are yellowish-brown, tubular tissues consisting of large, polygonal, cutinized cells; membranous cross walls occur at intervals.

Seed

Cross section.

Seed Coat. This consists of a layer of epidermal cells on the dorsal side, an epidermal layer and underlying parenchyma on the commissural side; the raphe is centrally situated in the latter tissue. The epidermal cells are tangentially elliptical except around the raphe where they are small and isodiametric or irregular. The outer walls of the epidermal cells are cutinized.

In surface view, the epidermal cells are polygonal in shape. An inner, thin cuticle is present.

Endosperm. This consists of polygonal cells with somewhat thick, colorless walls. The cell contents consist of aleurone grains and globules of fixed oil. There is a rosette crystal of calcium oxalate embedded in many of the larger aleurone grains.

Embryo. The embryo is about 1000 microns in length. It consists of radicle and 2 cotyledons, with the radicle lying toward the apex of the seed. The cells contain aleurone grains and globules of fixed oil.

Pedicel

The pedicel is similar to that described under fennel.

Carpophore

The carpophore is a modified extension of the pedicel and is divided.

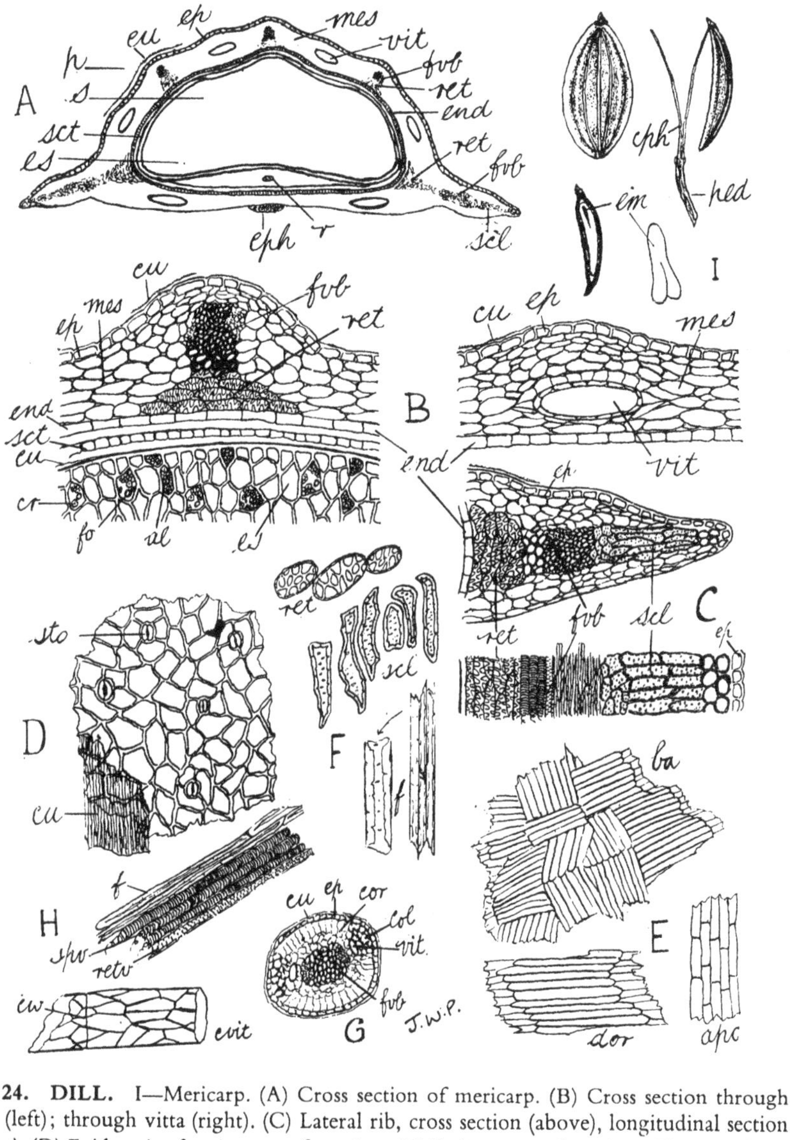

Fig. 24. DILL. I—Mericarp. (A) Cross section of mericarp. (B) Cross section through ridge (left); through vitta (right). (C) Lateral rib, cross section (above), longitudinal section (below). (D) Epidermis of pericarp, surface view. (E) Endocarp, surface view. (F) Reticulated cells, fibers and sclerenchyma occurring in lateral rib. (G) Cross section of pedicel. (H) Fibrovascular tissue; and commissural vitta (below).

al, aleurone; apc, apex, commissural side; ba, base; col, collenchyma; cor, cortex; cph, carpophore; cr, crystal; cu, cuticle; cw, cross wall; cvit, commissural vitta; dor, dorsal; em, embryo; end, endocarp; ep, epidermis; es, endosperm; f, fibers; fo, fixed oil; fvb, fibrovascular bundle; mes, mesocarp; p, pericarp; ped, pedicel; r, raphe; ret, reticulated cells; retv, reticulated vessel; s, seed; scl, sclerenchyma; sct, seed coat; spv, spiral vessel; sto, stoma; vit, vitta.

XXVII. FENNEL

Foeniculum vulgare Miller
Family: *Umbelliferae*

GENERAL

This is a biennial or perennial herb with finely divided green leaves. Flowers small, yellow, and produced in compound umbels. Fruits schizocarps; each mericarp contains a single seed. All parts of the plant are aromatic.

MORPHOLOGY

Schizocarp oblong-oval or elliptical, pale yellowish-brown with a tinge of green, usually with long pedicel, short stylopodium, 4 to 8 mm in length, and consisting of 2 mericarps attached to a divided carpophore.

Mericarp straight or curved, concave on the commissural side, convex on the dorsal side with 5 prominent longitudinal ridges or ribs.

The pericarp usually contains 4 dorsal and 2 commissural vittae but additional vittae are seen in some cross sections. The pericarp encloses a single seed with embryo embedded in the endosperm.

Fennel has a pleasant, aromatic odor and taste strongly resembling that of anise.

HISTOLOGY

Pericarp

Cross section.

Exocarp. This consists of thin cuticle and a layer of epidermal cells oblong-rectangular in shape or nearly so.

In surface view, the cells of the epidermis are polygonal in shape, and vary in size up to about 42 microns, long axis. Many small, rosette crystals of calcium oxalate occur in this tissue. Stomata are few, and surrounded by 3, 4, or 5 epidermal cells. The cuticle is smooth.

Mesocarp. This consists of tangentially elliptical or elongated parenchyma cells in the outer and inner mesocarp except where sclerenchymatous, reticulated cells flank the fibrovascular tissue.

Endocarp. This consists of a layer of thin-walled cells, tangentially elongated, narrow, except in the central commissural region where they are nearly quadrilateral in shape.

In surface view, the endocarp cells are transversely elongated, narrow, and vary in size up to about 90 microns, long axis, and up to 6 microns, short axis, except (a) in the central commissural region where the cells are longitudinally oblong-rectangular in shape, and vary in size up to about 68 microns, long axis, and to about 18 microns, short axis, and (b) in the basal commissural region where a number of transverse cells up to about 32 microns, long axis, and 9 microns, short axis, occur.

Fibrovascular Bundles. Five fibrovascular bundles traverse the outer mesocarp; 3 of these comprise the dorsal ridges and 2 the commissural ridges of the mericarp. The fibrovascular bundles are flanked on the outer side (above) by isodiametric, lignified, reticulated cells, and on the inner side (below) by tangentially elliptical, lignified, reticulated cells. Both spiral and reticulate vessels occur in the xylem tissue.

Vittae. There are usually 4 dorsal and 2 commissural vittae. Occasionally, however, branching of this tissue may result in 7, 8, or 9 vittae being seen in cross section. The vittae are yellowish-brown, elliptical, schizogenous, secretory cavities.

In surface view, the vittae are yellowish-brown, tubular tissues, sometimes branched, consisting of small, polygonal, cutinized cells; membranous cross walls occur at intervals.

Seed

Cross section.

Seed Coat. This consists of a layer of epidermal cells on the dorsal side, an epidermal layer and underlying parenchyma on the commissural side; the raphe is centrally situated in the latter tissue. The epidermal cells are tangentially oblong-rectangular; the parenchyma cells are tangentially elliptical except around the raphe where they are small and isodiametric or irregular in shape. The outer walls of the epidermal cells are cutinized.

In surface view, the epidermal cells are polygonal in shape. An inner, thin cuticle is present.

Endosperm. This consists of polygonal cells with moderately thick, colorless walls. The cell contents consist of aleurone grains and globules of fixed oil. There is a rosette crystal of calcium oxalate embedded in many of the larger aleurone grains.

Embryo. The embryo varies in length from about 1000 to 1300 microns, and consists of radicle and 2 cotyledons. It is embedded in the endosperm with the radicle toward the apex of the seed. The cells contain aleurone grains and globules of fixed oil.

Pedicel

Cross section.

Epidermis. This consists of a layer of cells oblong-rectangular in shape or nearly so, with moderately thick outer walls.

In surface view, the epidermal cells overlying the cortex collenchyma tissue are oblong-rectangular with somewhat thickened, pitted walls; there are no stomata. The epidermal cells overlying the cortex parenchyma tissue are polygonal in shape with thinner, faintly pitted walls; there are many stomata. The guard cells of the stomata are surrounded by 2, 3, or 4 epidermal cells with the first number predominant.

Cortex. This consists of parenchyma and collenchyma cells. The parenchyma cells immediately underlying the epidermis are palisade-like, the remainder are isodiametric or irregular in shape. The cell walls are thin. There are numerous chloroplasts in this tissue. The collenchyma cells fan out from the vittae to the epidermis. Soaking and bleaching is necessary to expand the tissues.

Vittae. There are 2 vittae; they are situated in the cortex.

Fibrovascular Tissue. A large, cylindrical mass of fibrovascular tissue forms the bulk and core of the pedicel.

Carpophore

The carpophore is a modified extension of the pedicel and is divided.

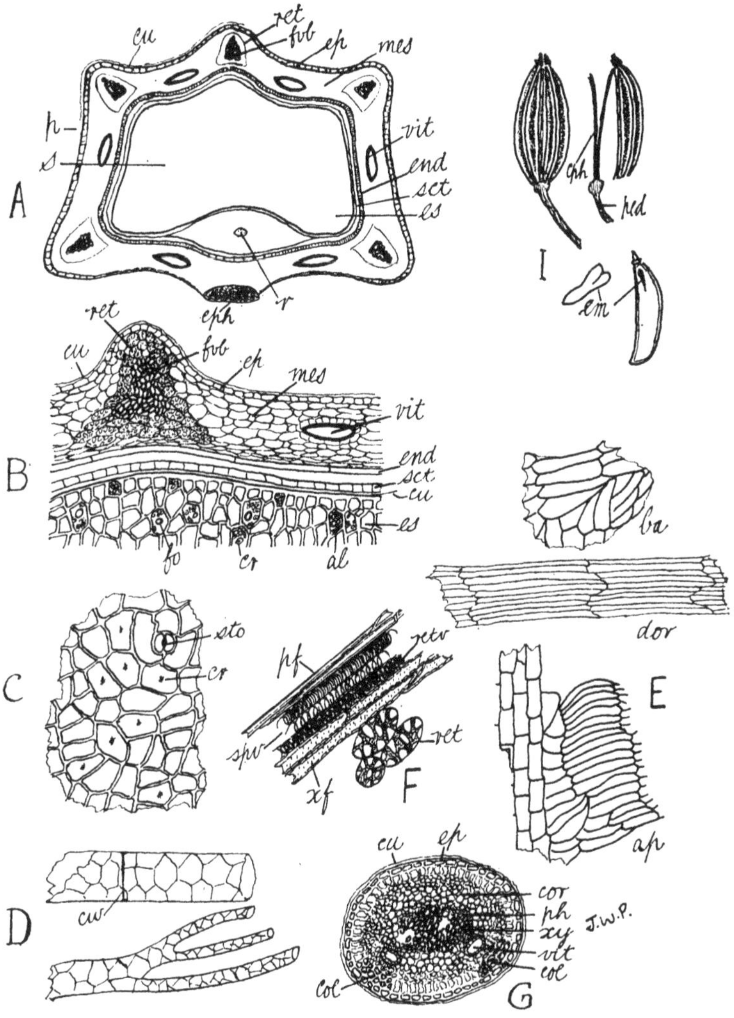

Fig. 25. FENNEL. I—Schizocarp (left); mericarp (right). (A) Cross section of mericarp. (B) Cross section through ridge. (C) Epidermis of pericarp, surface view. (D) Longitudinal view of vitta, simple and branched. (E) Endocarp, surface view. (F) Fibrovascular tissue and reticulated cells. (G) Cross section of pedicel.

al, aleurone; ap, apex; ba, base; col, collenchyma; cor, cortex; cph, carpophore; cr, crystal; cu, cuticle; cw, cross wall; dor, dorsal; em, embryo; end, endocarp; ep, epidermis; es, endosperm; fo, fixed oil; fvb, fibrovascular bundle; mes, mesocarp; p, pericarp; ped, pedicel; pf, phloem fibers; ph, phloem; r, raphe; ret, reticulated cells; retv, reticulated vessel; s, seed; sct, seed coat; spv, spiral vessel; sto, stoma; vit, vitta; xf, xylem fibers; xy, xylem.

Labiatae

GENERAL

The members of the Labiatae family included in this work are herbaceous or shrubby plants bearing simple, opposite leaves, and irregular, bilabiate flowers with 2 to 4 stamens and a deeply 4-loculed ovary; fruit of 4 achenes or nutlets.

XXVIII. SWEET MARJORAM

Majorana hortensis M.
Family: *Labiatae*

MORPHOLOGY

The commercial product consists of the dried leaves and floral parts of the plant; smaller leaves usually whole, larger leaves always broken. Leaves light green with a slight grayish tinge, obovate to broadly elliptical, margin entire, reaching about 21 mm in length, 11 mm in breadth; petiolate; net venation, pinnate; glandular punctate; pubescent. The dot-like oil glands and hairs visible with the aid of a good hand lens or the microscope.

Floral parts: Bract, light grayish-yellow, obovate, thin, reaching about 5.5 mm in length, 3.5 mm in breadth. Calyx gamosepalous, about 3 mm long, with 5 prominent teeth. Corolla sometimes present, gamopetalous, about 5 mm long; stamens 4, attached to corolla; style with bilobed stigma.

The odor of sweet marjoram is pleasant and aromatic; French marjoram is especially fragrant. The taste is warm, aromatic, slightly sharp, bitterish and somewhat camphoraceous.

HISTOLOGY

Leaf

Cross section. A thin cuticle covers the upper and lower epidermis. The cells of the upper and lower epidermis are nearly rectangular, the outer walls are thick, the inner walls are thin and arch inwardly. Small capitate hairs, oil glands sunken in depressions in the epidermis, and numerous thick-walled, papillose, pointed, nonglandular hairs, broad at the base, are present on both the upper and lower epidermis. The nonglandular hairs are uni- or multicellular with the latter type predominating. The multicellular hairs rarely exceed 4 cells, but an occasional hair with 5 cells occurs. The mesophyll consists of a single layer of palisade cells, and spongy tissue. Vascular tissue traverses the spongy mesophyll. The vascular bundle of the midrib is separated from the lower epidermis by 3 to 5 layers of collenchyma cells; a few collenchyma cells occur between the bundle and the upper epidermis.

In surface view, with the exception of the regions near and over the midrib and veins where the cells are more nearly rectangular and straight-walled, the cells of the upper and lower epidermis are irregular in shape and wavy-walled. The walls of the epidermal cells over the midrib are beaded, sometimes strongly; the walls of most of the other epidermal cells are faintly beaded. Stomata are few in the upper epidermis and abundant in the lower; the guard cells are surrounded by 2, 3, or 4 epidermal cells with the first number predominant. Nonglandular and glandular hairs occur on both sides of the leaf. The nonglandular hairs are described above; the capitate hairs (glandular) are small and have 1- or 2-celled stalks, they are not abundant. The oil glands are 12-celled and from about 65 to 75 microns in diameter.

Petiole

In cross section the petiole is somewhat crescent-shaped. The epidermal cells are rectangular or nearly so. Nonglandular and glandular hairs are present. Collenchyma tissue occurs in the corners, and a layer of collenchyma cells underlie the upper and lower epidermis. The vascular tissue occupies a central position in the mesophyll.

Flowering Tops

Bract. Cross section. The bract is thin and consists of (a) cuticle over each epidermis; (b) an upper epidermis of rectangular cells with outer walls about half as thick as those of the lower epidermis. A few stomata occur. Small capitate hairs are present but no oil glands or nonglandular hairs; (c) parenchymatous mesophyll tissue with chloroplasts; no palisade cells; (d) vascular tissue traversing the mesophyll; and (e) lower epidermis of rectangular cells with thick (7 to 10 microns) outer walls. Several stomata occur. Nonglandular hairs, capitate hairs, and oil glands are present. (Since the bract stands almost erect the lower is the outer epidermis, and the upper is the inner epidermis or that facing the calyx.)

In surface view, the epidermal cells of the bract are irregular and wavy-walled except over the midrib and at the base of the bract. Over the midrib the epidermal cells are oblong-rectangular and straight-walled; at the base they are vertically elongated and wavy-walled.

Calyx. The calyx is gamosepalous with 5 prominent teeth. The epidermal cells are irregular in shape and wavy-walled except at the base where they are oblong-rectangular and straighter-walled. On the outer epidermis, nonglandular and glandular hairs similar to those of the leaf occur. On the inner epidermis, small capitate hairs are present; the epidermal cells are mostly finely papillose, and numerous long, multicellular, papillose, pointed hairs arise from the throat region.

Corolla. The corolla is gamopetalous. The basal cells of the outer epidermis are vertically-elongated, wavy-walled between veins, nearly straight-walled over veins; the remaining epidermal cells are irregular in shape and wavywalled, papillose at the apexes of the lobes. The cells of the inner epidermis are similar to those of the outer but finely papillose from the higher basal region to the lobes. On the outer epidermis there are many short, uni- and multicellular papillose hairs, a few capitate hairs and oil glands. On the inner epidermis there are a few short, uni- and multicellular hairs, capitate hairs, and no oil glands.

Stamens and Pollen. The stamens are 4 in number. The epidermal cells of the anthers are papillose. The pollen grains are spherical but collapsed; about 20 to 22 microns in diameter.

Style. The style has a bilobe, papillose stigma.

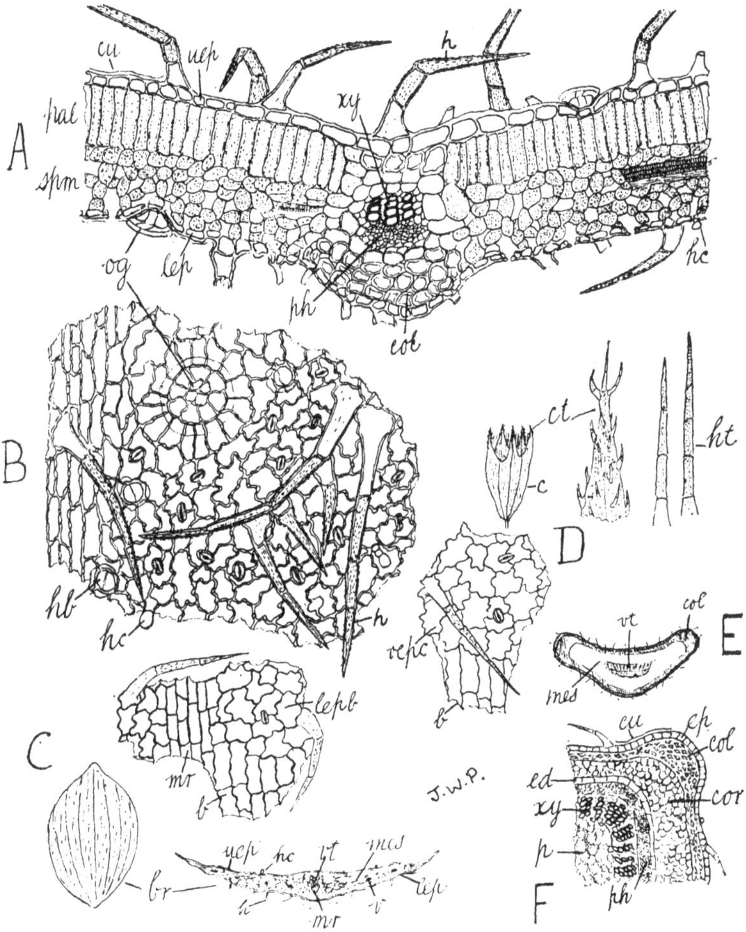

Fig. 26. SWEET MARJORAM. (A) Cross section through midrib. (B) Surface view of lower epidermis. (C) Bract; and cross section of bract (right); surface view of lower epidermis of bract (right, above). (D) Calyx; and surface view of outer epidermis of a calyx (below, left). (E) Cross section of petiole. (F) Cross section of corner of stem.

b, base; br, bract; c, calyx; col, collenchyma; cor, cortex; ct, calyx teeth; cu, cuticle; ed, endodermis; ep, epidermis; h, hair; hc, capitate hair; hb, hair base; ht, throat hairs; lep, lower epidermis; lepb, lower epidermis of bract; mes, mesophyll; mr, midrib; og, oil gland; oepc, outer epiderimis of calyx; pal, palisade cells; ph, phloem; p, pith; spm, spongy mesophyll; uep, upper epidermis; v, vein; vt, vascular tissue; xy, xylem.

XXIX. ORIGANUM

(Probably a variety of *Origanum vulgare*)
Family: *Labiatae*

[OREGANO, MEXICAN SAGE]

MORPHOLOGY

The commercial product consists of the dried leaves and floral parts of the plant; smaller leaves usually whole, larger leaves always broken. Leaves light green to greenish-brown, blade ovate, margin crenate to crenate-serrate, about 3 cm in length; petiolate; net venation, pinnate; glandular punctate; pubescent. The dot-like oil glands and hairs visible with the aid of a hand lens or the microscope.

Flowers crowded into spike-like clusters, bracts small, broadly ovate, imbricated, 2.5 to 3 mm long, 2 mm broad, hairs on the outer surface visible to the naked eye, hairs on the inner surface visible only with the microscope. Calyx small, gamosepalous, 4-toothed, deeply cleft on 2 sides, without throat hairs. Corolla much longer than the calyx, about 5 mm, gamopetalous; stamens 4, attached to corolla; style plumose.

Origanum has a strong, camphoraceous odor and a warm, pungent, bitter, camphoraceous taste.

HISTOLOGY

Leaf

Cross section. A thin cuticle covers the upper and lower epidermis. The epidermal cells are nearly rectangular except around the base of hairs where they are angular. The outer cell walls of the upper epidermis are thick. On both the upper and lower epidermis there are numerous curved, pointed, unicellular, nonglandular hairs, some papillose, others smooth; unicellular oil glands with very short unicellular stalks; very small, unicellular capitate hairs with multicellular stalks; and small, bicellular, biseriate glandular hairs with unicellular stalks. The mesophyll consists of a single layer of palisade cells, and spongy tissue. The vascular tissue traverses the spongy mesophyll. Collenchyma tissue separates the vascular bundle of the midrib and veins from the upper and lower epidermis. The midrib is large, and the veins are pronounced.

In surface view, the cells of the upper and lower epidermis are irregular, angular, or polygonal in shape, except over the midrib and veins where they are oblong-rectangular or nearly so. The cell walls are uneven in thickness, occasionally beaded near the midrib, veins and hairs, and strongly beaded over the midrib and veins. The nonglandular hairs vary in length to about 900 microns; the oil glands are about 40 microns in diameter; the gland cells of the biseriate hairs are elongated to about 32 microns; and the small capitate hairs have a glandular head about 22 microns in diameter. There are stomata on both sides of the leaf, sparse on the upper surface, numerous on the lower; the guard cells are surrounded by 2, 3, or 4 epidermal cells.

Petiole

In cross section, the petiole is somewhat crown-shaped. The cells of the epidermis are nearly rectangular. Numerous nonglandular and glandular hairs are present. Collenchyma tissue occupies the corners and underlies the upper and lower epidermis. On the sides of the petiole, parenchymatous palisade-like cells with chloroplasts underlie the epidermis. The mesophyll cells are isodiametric, oval, or elliptical in shape. The main vascular tissue occupies a central position in the mesophyll, and a small vascular bundle occurs in each upper corner of the petiole.

Flowering Tops

Bract. The epidermal cells of the bract are similar in shape and arrangement to those of the leaf. Nonglandular and glandular hairs occur. The nonglandular hairs are numerous on the outer epidermis, particularly at the margins; few on the inner. Stomata are present and they are similar to those of the leaf.

Calyx. The calyx is gamosepalous, small, 4-toothed, and deeply cleft on 2 sides. The cells and hairs of the outer epidermis are similar to those of the leaf and bract; there are no hairs on the inner epidermis.

Corolla. The corolla is gamopetalous; the corolla tube is long, the lobes are regular. The cells of the inner epidermis are irregular in shape; irregular and papillose toward and at the lobes. The cells of the outer epidermis of the corolla tube are oblong-rectangular in shape with straight, beaded walls; those of the outer epidermis of the lobes are irregular in shape with "looped" walls. Numerous nonglandular and glandular hairs occur on the outer epidermis of that portion of the corolla which extends beyond the calyx.

Stamens. The stamens are 4 in number and arise from the throat of the corolla. The outer epidermal cells of the anthers are papillose. The epidermal cells of the filaments are oblong-rectangular.

Pistil. The pistil consists of a 2-loculed ovary, style, and bilobed, spinose stigma. The bilobed nature of the stigma is obscured in a mass of prominent, short, spinelike cells among which remains of pollen grains are usually present. The pollen grains are spherical or nearly so, 3-ridged, and about 22 microns in diameter.

Stem

Cross section. The stem is square or nearly so. The cells of the epidermis are nearly rectangular; angular where hairs arise. Numerous hairs are present. There are about 3 layers of collenchyma cells in each corner, and 1 layer underlying the epidermis on each side of the stem. A broken ring of fibers and a few stone cells occur in the cortex. With the exception of about 3 layers of cells adjoining the phloem tissue, the cells of the cortex have brown contents. The fibrovascular tissue forms an unbroken ring around the pith. The cells of the pith are isodiametric; most are pitted.

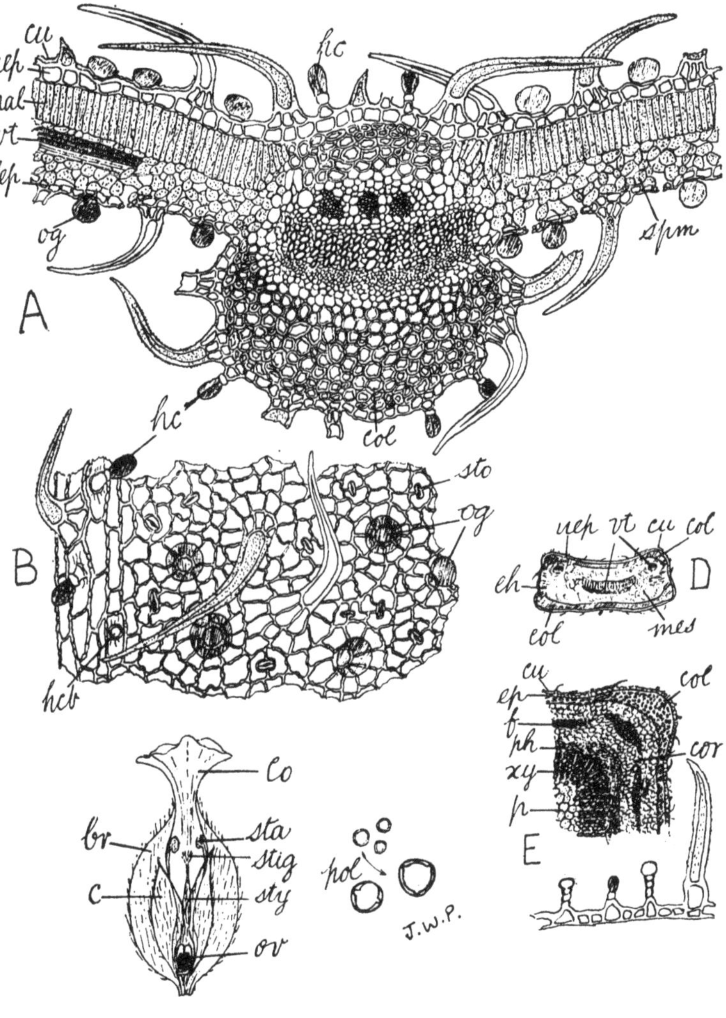

Fig. 27. ORIGANUM. (A) Cross section through midrib. (B) Lower epidermis, surface view. (C) Flower. (D) Cross section of petiole. (E) Cross section of corner of stem; below, stem hairs.

br, bract; c, calyx; ch, chloroplast cells; co, corolla; col, collenchyma; cor, cortex; cu, cuticle; ep, epidermis; f, fibers; hc, capitate hair; hcb, capitate hair base; lep, lower epidermis; mes, mesophyll; og, oil gland; ov, ovary; pal, palisade cells; ph, phloem; p, pith; pol, pollen; spm, spongy mesophyll; sta, stamen; stig, stigma; sto, stoma; sty, style; uep, upper epidermis; vt, vascular tissue; xy, xylem.

XXX. PEPPERMINT

Mentha piperita L.
Family: *Labiatae*

MORPHOLOGY

Dried leaves, whole, broken, crumpled, or shrivelled; dark green; mature leaf blade commonly about 45 mm in length, and about 23 mm in width; petiolate, ovate, margin serrate, apex acute, midrib prominent; net venation, pinnate; glandular punctate, the dot-like oil glands visible with hand lens; non glandular hairs present on the lower epidermis, occurring on the midrib, veins and at margin; a few nonglandular hairs on the upper epidermis, occurring on the midrib; hairs visible with the aid of a hand lens.

Floral parts: Calyx gamosepalous, purplish, small, about 3 mm long, glandular punctate, equally 5-toothed, teeth prominent and long, hairs arising from the calyx teeth but throat naked. Corolla sometime present, gamopetalous, stamens 4, style with bilobed stigma.

Odor characteristic, sweetish, strong; taste aromatic, warm, pungent, with a cooling aftertaste.

HISTOLOGY

Leaf

Cross section. A cuticle covers the upper and lower epidermis. The cells of the upper epidermis are nearly rectangular in shape, regular in arrangement, and have moderately thick outer walls. The cells of the lower epidermis are nearly rectangular, less regular, and have somewhat thinner walls than those of the upper epidermis. Oil glands, capitate hairs, uni- and multicellular nonglandular hairs occur on the lower surface; the nonglandular hairs arise from the epidermis over the midrib, veins, and at the margin of the leaf blade.

A few nonglandular hairs occur on the upper surface over the midrib. Nonglandular hairs do not occur on other parts of the leaf blade. The mesophyll consists of 1 layer of palisade cells, and spongy tissue. Vascular tissue traverses the spongy mesophyll. Between the lower epidermis and the vascular tissue of the midrib and veins there is a conspicuous group of collenchyma cells giving support, bulk and prominence to these parts.

In surface view, the cuticle is striated over the midrib, veins, and at the margins of the leaf; it is otherwise clear. With the exception of the regions near and over the midrib and veins, the cells of the upper epidermis are irregular in shape with wavy walls, and vary in size up to about 75 microns, long axis, with some irregular-elongated cells near the midrib reaching 90 microns. Near and over the midrib and veins the cells become oblong-rectangular or nearly so, with nearly straight, infrequently beaded walls. At the margins the cells become smaller, polygonal, with straight walls, some beaded. Over the teeth and at the apex the cell walls are slightly wavy. Stomata do not occur on the upper surface. The cells of the lower epidermis are similar to those of the upper but somewhat smaller. Stomata are numerous on the lower surface; the guard cells are surrounded by 2 epidermal cells. Numerous small crystals are contained in the cells of the upper and lower epidermis; they are more clearly observed with polarized light. The nonglandular hairs arising from the lower epidermis over the midrib are long, multicellular, commonly 5 or 6 cells, thick-walled, papillose, and pointed; those on the veins and at the margin are similar but shorter and with fewer cells, many having only 1 or 2 cells. The few hairs arising from the upper epidermis over the midrib are multicellular, papillose, but shorter and less robust than those of the lower surface. Capitate hairs and oil glands occur on both the upper and lower epidermis. The capitate hairs are small and of two kinds: those with an oval or spherical head, 15 to 20 microns in diameter, and a short, 1-celled stalk; and those with a head only slightly enlarged, and a longer, 2-celled stalk. The oil glands are large, about 65 to 75 microns in diameter; most have 8 cells, but glands with 10 or 12 cells occur. With polarized light, a crystal group is seen in many of these glands.

Petiole

In cross section, the petiole is somewhat crescent-shaped, the upper side concave. A thin cuticle covers the epidermis. The epidermal cells are rectangular and regular. Long, multicellular, nonglandular hairs are present. Collenchyma tissue occupies the corners and underlies the upper and lower epidermis. The mesophyll cells are isodiametric with moderately thick walls.

The main vascular tissue is central in the mesophyll, and a small vascular bundle is situated in each corner adjoining the collenchyma tissue.

Flowering Tops

Calyx. The calyx is gamosepalous with 5 prominent teeth. The cells of the outer and inner epidermis are irregular in shape and wavy-walled, except toward and at the base of the calyx tube where they are oblong-rectangular and straight-walled, or nearly so. Capitate hairs and oil glands occur on the outer epidermis. Prominent, uni- and multicellular, papillose, pointed, non-glandular hairs arise from the calyx teeth. There are no hairs of any kind on the inner epidermis. Vascular tissue traverses the mesophyll, 3 bundles merging in each tooth.

Corolla. The corolla is gamopetalous. The cells of the outer and inner epidermis are irregular in shape and papillose at the lobes; irregular in shape, wavy-walled, and finely papillose below the lobes; oblong-rectangular and wavy-walled in the region of the stamens; and oblong-rectangular and straight-walled toward and at the base of the corolla. On the outer epidermis there are capitate hairs, oil glands, and a few 1- or 2-celled, papillose, non-glandular hairs, all occurring on the lobes. There are no hairs of any kind on the inner epidermis. Vascular tissue traverses the mesophyll.

Stamens. The stamens are 4 in number, short, and similar. Crystals occur in the epidermal cells of the anthers. The pollen grains are spheroidal, and from about 25 to 35 microns in diameter.

Style and Stigma. The epidermal cells of the style are oblong-rectangular, and contain crystals. The stigma is bilobed and papillose.

Stem

Cross section. The stem is square. A thin cuticle covers the epidermis. The epidermal cells are nearly rectangular and regular. Occasional hairs occur. A group of collenchyma cells occupies and bulges each corner; a layer of collenchyma cells underlies the epidermis on all sides. The cells of the cortex are elliptical or oval with small intercellular spaces. An endodermis is present. A broken ring of fibers and fibrovascular tissue surrounds the pith. The pith consists of isodiametric, oval, or elliptical parenchyma cells with small intercellular spaces.

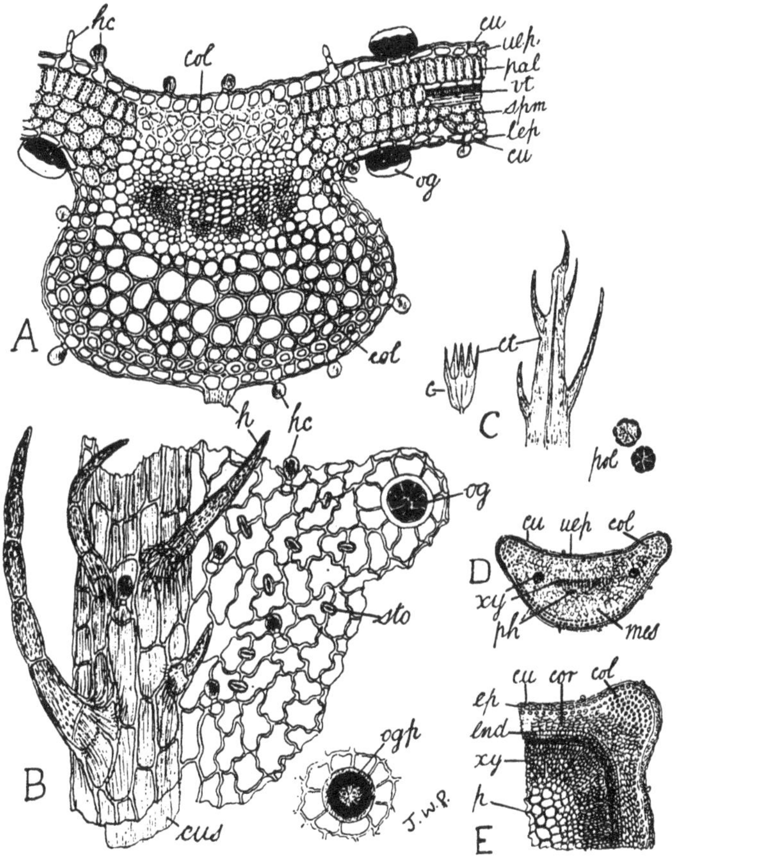

Fig. 28. PEPPERMINT. (A) Cross section through midrib. (B) Lower epidermis, surface view. (C) Calyx and calyx tooth. (D) Cross section of petiole. (E) Cross section of corner of stem.

c, calyx; col, collenchyma; cor, cortex; ct, calyx tooth; cu, cuticle; cus, cuticle, striated; end, endodermis; ep, epidermis; h, hair; hc, capitate hair; lep, lower epidermis; mes, mesophyll; og, oil gland; ogp, oil gland with crystals, polarized light; pal, palisade cells; p, pith; ph, phloem; pol, pollen; spm, spongy mesophyll; sto, stoma; uep, upper epidermis; vt, vascular tissue; xy, xylem.

XXXI. SPEARMINT

Mentha spicata L.
Family: *Labiatae*

Fig. 29. PEPPERMINT (*left*), **SPEARMINT** (*right*).

MORPHOLOGY

Dried leaves, whole, broken, crumpled, or shrivelled; olive green; mature leaf blade commonly about 35 mm long, and about 13 mm wide; petiole very short, about 2 to 4 mm long; blade lance-ovate to elliptical, margin unequally serrate, apex acute, midrib prominent; net venation, pinnate; glandular punctate, the dot-like oil glands visible with hand lens; a few nonglandular hairs on the lower surface of the leaf blade arising from the epidermis over the midrib, some occurring at the margin near the teeth; hairs not visible without the aid of the microscope; no nonglandular hairs on the upper surface.

Floral parts similar to those of peppermint.

Odor and taste similar to that of peppermint but to a leasser degree; no cooling aftertaste.

HISTOLOGY

The histological structure of spearmint is similar to that of peppermint. The chief differences, and these are probably more morphological than histological, lie in the leaves and are (a) the absence of nonglandular hairs on the upper epidermis; (b) fewer, and shorter, nonglandular hairs on the lower surface over the midrib, with few, if any, over the veins; and (c) the absence of crystals in the oil glands.

XXXII. ROSEMARY

Rosmarinus officinalis L.
Family: *Labiatae*

MORPHOLOGY

Dried leaves, brownish-green, sessile, linear, slightly curved, margins revolute, from about 6 to 25 mm long, midrib prominent, glandular-punctate, lower surface wooly except over the midrib but concealed by marginal curling of the leaf.

No floral parts are present.

Rosemary leaves have a tea-like fragrance. When crushed they yield an agreeable, aromatic odor with a slight camphoraceous note. The taste is aromatic, pungent, somewhat bitter and slightly camphoraceous.

HISTOLOGY

Leaf

Cross section. The cells of the upper epidermis are compressed, small, and inconspicuous; cuticle wrinkled. Unless the section has been well soaked in water, or the cells expanded with the aid of bleaching solution, it is possible to overlook the existence of the upper epidermis in cross section. Remains of occasional hairs occur on the upper epidermis. Collenchyma tissue forms a thick-walled hypodermal layer underlying the upper epidermis, and wedge-shaped pockets of collenchyma cells separate groups of mesophyll cells. The palisade cells are up to about 80 microns, long axis, and form 1 to 3 layers. Spongy tissue completes the mesophyll. A dense, tangled mass of branched, multicellular, smooth hairs covers the lower epidermis. The branches of the hairs are curved and pointed. Some hairs give rise to a bulbous, glandular branch. Capitate hairs and 8-celled oil glands are also present. The lower

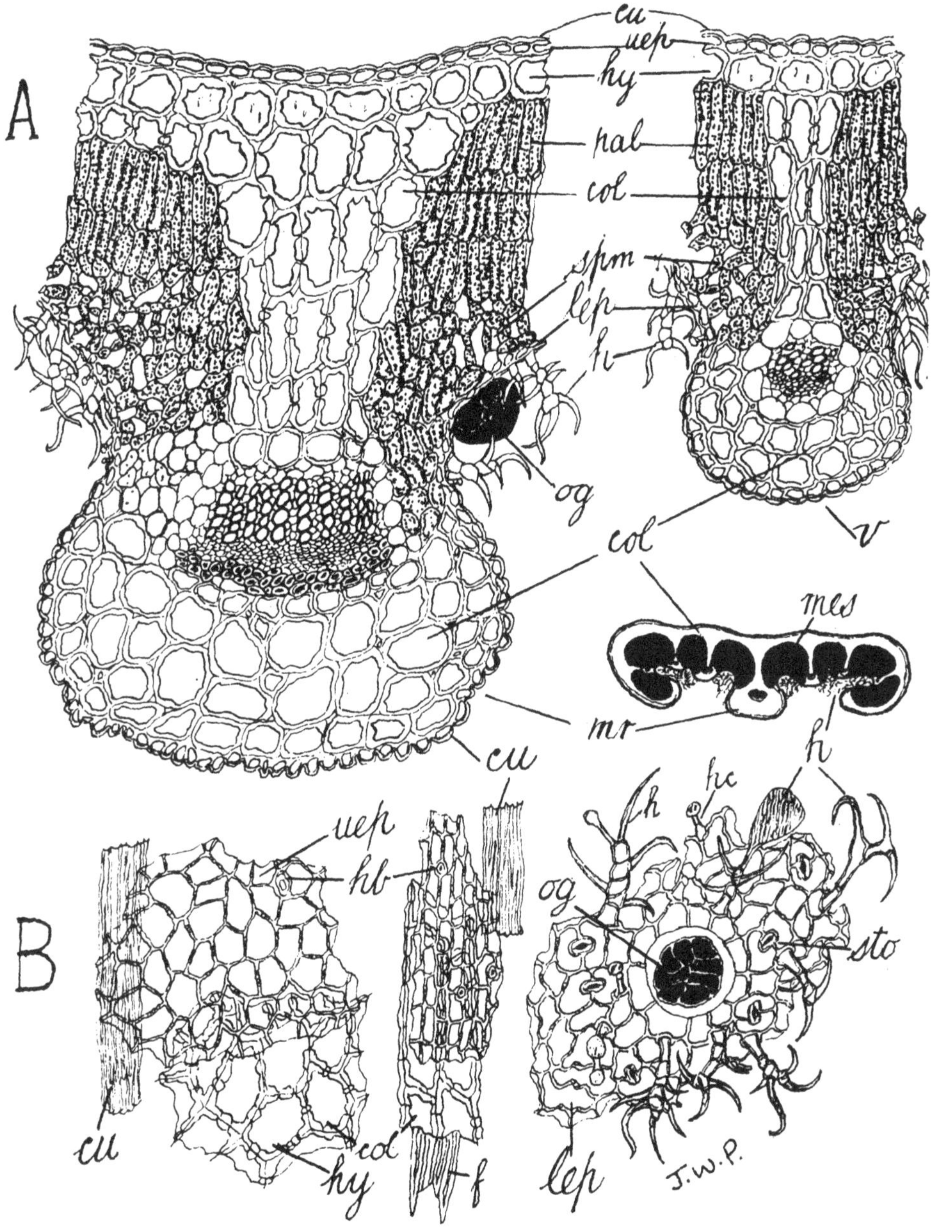

Fig. 30. ROSEMARY. (A) Cross section through midrib (right); through vein (far right); cross section of leaf (below). (B) Upper epidermis (right); lower epidermis of midrib (center); lower epidermis (far right); all in surface view.

col, collenchyma; cu, cuticle; f, fibers; h, hair; hc, capitate hair; hb, hair base; hy, hypodermis; lep, lower epidermis; mes, mesophyll; mr, midrib; og, oil gland; pal, palisade cells; spm, spongy mesophyll; sto, stoma; uep, upper epidermis; v, vein.

epidermis and its cuticle are closely wrinkled at the midrib. The midrib is prominent; the fibrovascular bundle is separated from the hypodermis by a wedge-shaped group of collenchyma cells, and from the lower epidermis by 3 or 4 layers of collenchyma cells. The vessels are arranged in radial rows. Thick-walled fibers adjoin the phloem tissue. The vascular tissue of the veins is surrounded by a parenchymatous sheaf, and separated from the hypodermis by a wedge-shaped group of collenchyma cells, and from the lower epidermis by isodiametric or irregular-shaped collenchyma cells.

In surface view, the cells of the upper epidermis are polygonal in shape; the walls are of uneven thickness and beaded. There are no stomata. The collenchyma cells are large, up to about 110 microns, long axis, and polygonal in shape with thick, beaded walls. The cells of the lower epidermis are similar in shape to those of the upper but slightly larger; the walls are somewhat thinner and not so clearly beaded. The epidermal cells over the midrib are oblong-rectangular in shape and straight-walled. Stomata are frequent on the lower surface; the guard cells are surrounded by 2 epidermal cells.

Stem

Only occasional pieces of stem are present. The anatomy of the stem is similar to that of other labiate plants described in this work. An endodermis consisting of cells averaging about 10 microns in width, and about 26 microns, radial axis, is present. The vascular tissue of the stem is abundant and forms a complete ring around the pith. The pith cells have thick, pitted walls.

XXXIII. SAGE

Salvia officinalis L.
Family: *Labiatae*

MORPHOLOGY

Sage, Domestic

Leaves grayish-green, oblong-lanceolate, reaching about 8.5 cm in length, 2.5 cm in breadth, petiolate, apex acute, margin finely crenate; net venation, pinnate-reticulate; densely pubescent but not wooly; soft and velvety in texture; glandular-punctate, but glands obscured by hairs, visible only with the microscope.

Odor aromatic, strongly so when the leaves are crushed; taste aromatic and somewhat bitter.

Sage, Dalmatian

Dried leaves, whole, broken, or crushed; gray, tinged with green; wide variation in size; apex acute or obtuse; densely pubescent, wooly; otherwise as Domestic sage.

Odor strong, fragrant and aromatic; taste aromatic, warm, somewhat astringent and a little bitter.

Sage, Cyprus

Dried leaves, grayish-green to dark gray; hard, brittle; pubescent but not wooly; otherwise as Dalmatian sage.

Odor fragrant, aromatic, and somewhat camphoraceous; taste faintly aromatic, a little bitter, and somewhat camphoraceous.

Sage, Spanish

Dried leaves, dark gray; whole leaves usually small; not so wooly as Dalmatian sage, but otherwise as that product.

Odor aromatic with a camphoraceous note; taste warm, a little bitter, somewhat astringent and camphoraceous.

HISTOLOGY

Sage, Dalmatian

Leaf. Cross section. A cuticle, well-developed over the midrib, covers the epidermis. The cells of the upper and lower epidermis are rectangular or nearly so, but somewhat irregular over the midrib. Numerous nonglandular hairs, capitate hairs, and oil glands occur on both surfaces. The nonglandular hairs are uni- or multicellular with the latter type predominating. The multicellular hairs are very long, narrow, pointed, mostly smooth but some papillose, clear or with dark contents, and form a tangled mass. The capitate hairs have a bicellular, biseriate head and unicellular stalk, or a unicellular head and multicellular stalk. The midrib is prominent; its vascular bundle is large, and separated from the upper and lower epidermis by parenchyma and well-developed collenchyma tissue. Collenchyma is also present above and below the vascular tissue of the veins but to a lesser degree. The vascular tissue of the veins is surrounded by a parenchymatous sheaf. The mesophyll consists of from 1 to 3 layers of palisade cells, and spongy tissue.

In surface view, the cells of the upper epidermis are polygonal or irregular in shape with nearly straight, moderately thick, beaded walls. A few stomata are present. The cells of the lower epidermis are irregular in shape with nearly straight, or somewhat wavy walls, some beaded; the walls are slightly thinner than those of the upper epidermis. Stomata are numerous; the guard cells are surrounded by 2 or 3 epidermal cells, the former predominating. The oil glands have 8 to 12 cells. A study of the oil glands of Dalmatian sage is rendered difficult by the tangled mass of hairs surrounding them. The walls of the cells of the glands are very thin and easily obscured by the density of the oil. The nature of the glands is better seen in Spanish sage where the hairs are much shorter and the glands more completely visible. Clearing the oil with the aid of alcohol or other solvent is not much help; a weak, about 10 per cent, bleaching solution is better. Sections should be left in this solution for several hours or overnight. This treatment will generally clear the glands making a study of the thin walls of their cells possible.

Petiole. Cross section. The epidermal cells are rectangular or nearly so. Numerous nonglandular and glandular hairs are present. Collenchyma tissue occupies the corners and underlies the epidermis. The cells of the mesophyll are isodiametric, and vary in size to about 50 microns. The main fibrovas-

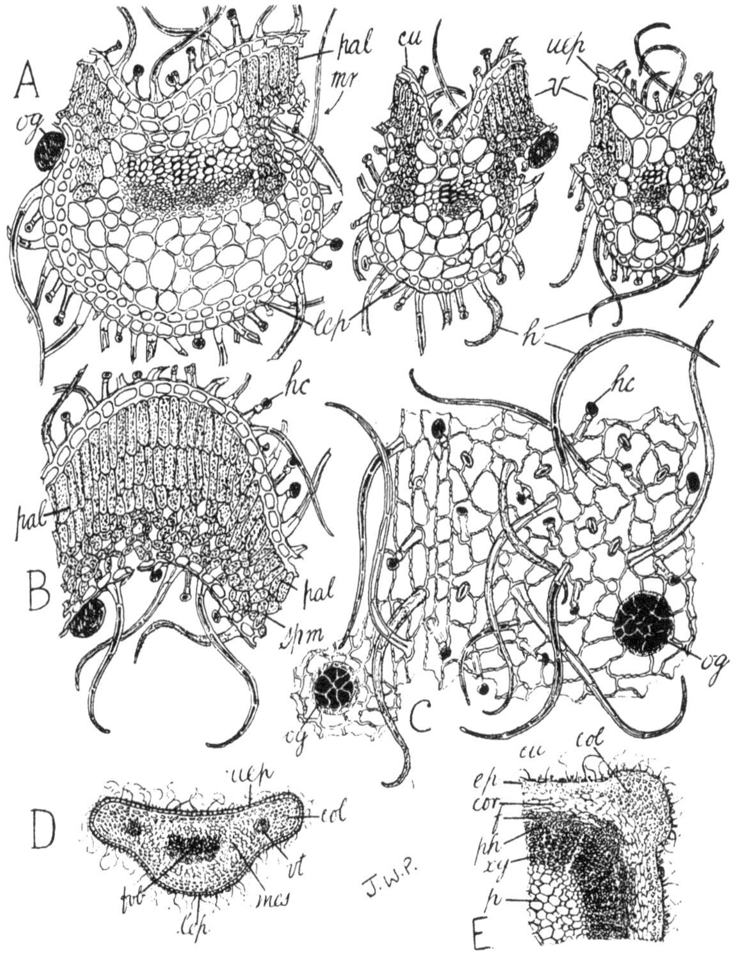

Fig. 31. SAGE. (A) Cross section through midrib; cross sections of a larger and a smaller vein at right. (B) Cross section of a portion of leaf blade. (C) Lower epidermis, surface view; (D) Cross section of petiole. (E) Cross section of corner of stem.

col, collenchyma; cor, cortex; cu, cuticle; ep, epidermis; f, fibers; fvb, fibrovascular bundle; h, hair; hc, capitate hair; lep, lower epidermis; mes, mesophyll; mr, midrib; og, oil gland; pal, palisade cells; p, pith; ph, phloem; spm, spongy mesophyll; uep, upper epidermis; v, vein; vt, vascular tissue; xy, xylem.

cular bundle is large, and occupies a central position in the mesophyll. A small fibrovascular bundle adjoins the collenchyma tissue in each corner.

Flowering Tops. Floral parts are not present in the product.

Stem. Cross section. The stem is square. The epidermal cells are rectangular or nearly so. Numerous nonglandular and glandular hairs are present. Collenchyma tissue occupies the 4 corners and underlies the epidermis on all sides of the stem. The cortex consists of thin-walled parenchyma cells. A broken ring of fibers, and an unbroken ring of fibrovascular tissue surrounds the pith. The cells of the pith are isodiametric, thin-walled, and up to about 110 microns in diameter.

Sage, Cyprus

The histological structure of Cyprus sage is similar to that of Dalmatian sage but the nonglandular hairs are fewer, somewhat thicker, and much shorter. Many of the capitate hairs are longer than those of Dalmatian sage.

Sage, Spanish

Spanish sage is anatomically similar to the Dalmatian product but the nonglandular hairs are shorter. The capitate hairs are fewer in number.

XXXIV. SAVORY

Satureia hortensis L.
Family: *Labiatae*

MORPHOLOGY

Dried leaves, greenish-brown, small, reaching about 10 mm in length; shape oblanceolate with acute tip, or spatulate with obtuse tip; tapering to short petiole; mostly folded upwards along the midrib; margin entire; net venation, pinnate; glandular punctate; pubescent. The dot-like oil glands and the hairs are visible with hand lens.

Floral parts: Calyx gamosepalous, 3 to 4 mm long; upper lobe with 3 teeth, short, united; lower lobe with 2 teeth, awl-like, deeply cleft; throat hairy. Corolla sometimes present, gamopetalous, about 4 to 5 mm long; stamens 4, didynamous; style with bilobed stigma.

Odor fragrant and aromatic; taste aromatic, warm, slightly sharp and somewhat camphoraceous.

HISTOLOGY

Leaf

Cross section. A thin cuticle covers the epidermis. The cells of the upper and lower epidermis are rectangular in shape or nearly so. Nonglandular hairs, capitate hairs; and oil glands are present. The midrib is not prominent. The fibrovascular bundle of the midrib is surrounded by a parenchymatous sheaf, and separated from the upper epidermis by parenchyma and collenchyma cells, and from the lower epidermis by parenchyma cells. The mesophyll consists of 2 layers of palisade cells, and spongy tissue with small intercellular spaces. It should be observed that the cells of the spongy mesophyll closely resemble those of the palisade mesophyll, and are very closely packed.

In surface view, the cells of the upper and lower epidermis are irregular in shape, and vary in size to about 88 microns, long axis; the cell walls are beaded. Uni- and multicellular, papillose, pointed hairs with dark contents, capitate hairs, and oil glands, occur on both sides of the leaf but are more numerous on the upper surface. The capitate hairs have a unicellular stalk. The oil glands are 12-celled, large, from about 88 to 96 microns in diameter, and are sunken in depressions in the epidermis. Stomata occur on both surfaces of the leaf; the guard cells are surrounded by 2 epidermal cells.

Petiole

In cross section, the petiole is somewhat crescent-shaped. The epidermal cells are rectangular or nearly so; hairs are present. Collenchyma cells occupy the corners, and a layer of collenchyma cells underlie the upper and lower epidermis. The cells of the mesophyll are isodiametric, oval, or elliptical, and up to about 44 microns, diameter or long axis. The fibrovascular tissue occupies a central position in the mesophyll.

Flowering Tops

Calyx. The calyx is gamosepalous, 5-toothed, and large. Two of the five teeth are prominent, long, and deeply cleft; the remaining three are short and shallowly cleft, united. Stiff, pointed hairs occur on the teeth. Nonglandular hairs, capitate hairs, and oil glands are present on the outer epidermis. Numerous, long, multicellular, papillose, pointed hairs arise from the calyx throat. The cells of the inner epidermis in the region of the throat hairs are finely papillose.

Corolla. The corolla is gamopetalous. In surface view, the outer and inner epidermal cells are oblong-rectangular in shape and straight-walled in the basal region of the corolla tube, oblong-rectangular or irregular with somewhat wavy walls in the lower lobal region, irregular and wavy-walled in the lobal region, and irregular and papillose at the apexes of the lobes. Small, prismatic crystals are seen in the cells of the lobes when examined with polarized light. Nonglandular hairs, capitate hairs, and oil glands occur on the outer epidermis of the corolla where it extends beyond the calyx. There are no hairs of any kind on the inner epidermis.

Stamens and Style. The stamens are 4 in number and didynamous. The pollen grains are spheroidal and vary in size from about 28 to 42 microns; they are granular and have 6 furrows. The style has a bilobed, papillose stigma.

Stem

Cross section. The stem is square or nearly so. Hairs similar to those

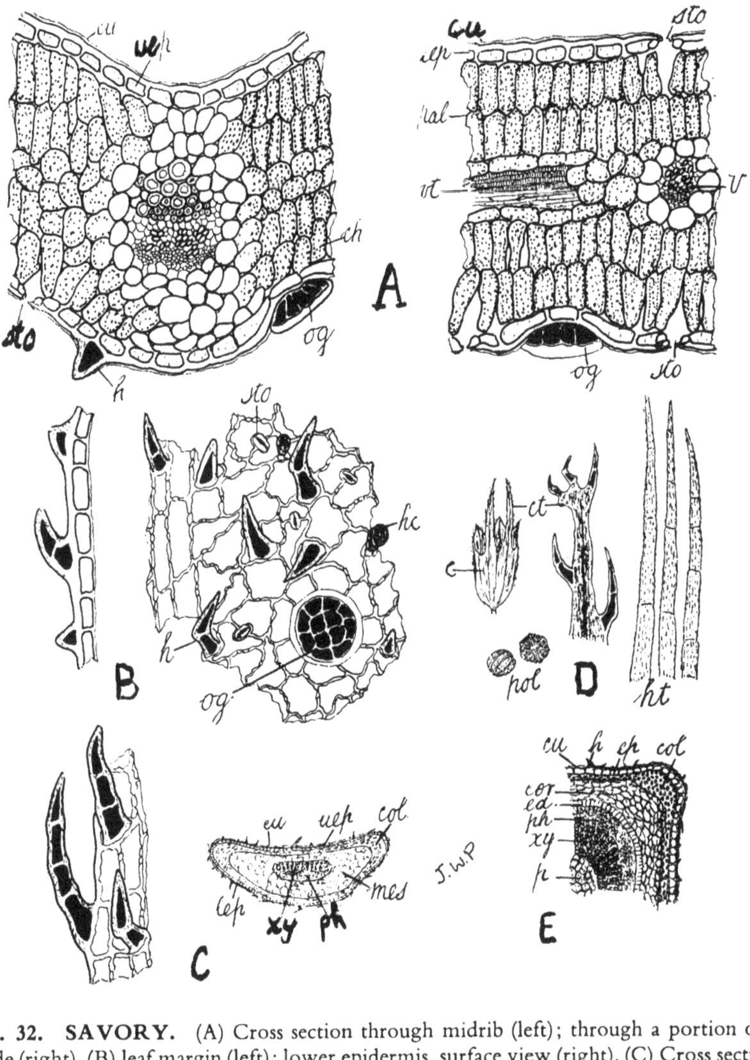

Fig. 32. SAVORY. (A) Cross section through midrib (left); through a portion of leaf blade (right). (B) leaf margin (left); lower epidermis, surface view (right). (C) Cross section of petiole (above, right); surface view of epidermis of petiole (left). (D) Calyx, calyx tooth, and throat hairs. (E) Cross section of corner of stem.

c, calyx; ch, chloroplasts; col, collenchyma; cor, cortex; ct, calyx tooth; cu, cuticle; ed, endodermis; ep, epidermis; h, hair; hc, capitate hair; ht, throat hairs; lep, lower epidermis; mes, mesophyll; og, oil gland; pal, palisade cells; p, pith; ph, phloem; pol, pollen; sto, stoma; uep, upper epidermis; v, vein; vt, vascular tissue; xy, xylem.

already described occur on the epidermis. The epidermal cells are rectangular or nearly so. Collenchyma cells occupy the corners, and a layer of collenchyma tissue underlies the epidermis on all sides of the stem. An endodermis separates the cortex from the vascular tissue. The fibrovascular tissue forms an unbroken ring around the pith. The pith consists of pitted cells up to about 88 microns in diameter.

XXXV. SWEET BASIL

Ocimum basilicum L.
Family: *Labiatae*

GENERAL

Several varieties of this herb find their way into commerce, including the small leaves of the common basil, the larger leaves of the purple leaf basil, the Italian basil, and the lettuce leaf basil. The large lettuce leaf has a mild aroma while the common basil with its very small leaf is very pungent. The larger leafed varieties are more commonly cultivated in the United States of America.

American producers of sweet basil prepare the commercial product by dehydrating the fresh leaves with controlled artificial heat and air. This process is carefully carried out to preserve color and aroma. The leaves may also be dehydrated by drying them in the sun but with this method precautions must be taken against night moisture. Fragmentation of the dried leaves is accomplished mechanically. A series of machines mill, brush, shake, blow and screen the leaf particles through specified size aperatures. The dried, fragmented herb is graded fine, medium, and coarse, enabling spice merchants to offer selective sizes suitable to the needs of their customers and consumers.

MORPHOLOGY

Leaves ovate, tip acute, petiolate; green; small leaves from about 1 to 5 cm, large leaves reaching about 10 cm in length; margin of small leaves finely serrate, that of large leaves indistincly serrate; net venation, pinnate; glandular punctate; slightly pubescent. The dot-like oil glands and the hairs are visible with a hand lens.

Floral parts: Calyx large, reaching about 7 mm, gamosepalous, irregular, and consisting of (a) lower lobe with twin awl-like teeth; (b) 2 upper lobes, deeply cleft, each with a short tooth; and (c) a large, broadly cordate labiate lobe. Calyx throat and inner surface of labiate lobe hairy. Corolla sometimes present; gamopetalous, longer than calyx; stamens 4, didynamous; style with bilobed stigma.

The dried leaves of the large leafed varieties of sweet basil have an aromatic, fragrant and sweet odor faintly reminiscent of anise. The taste is aromatic, warm, and mildly pungent. The dried leaves of the common basil have a less fragrant odor and are more pungent.

HISTOLOGY

Leaf

Cross section. A thin cuticle covers the epidermis on both sides of the leaf. The cells of the upper and lower epidermis are similar and nearly rectangular except where hairs and glands arise. The outer walls of the cells of the upper epidermis are thicker than those of the lower epidermal cells. Nonglandular, pointed hairs, capitate hairs, and oil glands occur on both surfaces of the leaf. A single palisade layer underlies the upper epidermis. The palisade cells are up to about 88 microns, long axis, and have an average width of about 22 microns. Spongy tissue completes the mesophyll. The vascular tissues traverse the spongy mesophyll.

The vascular tissue of the midrib and large veins is separated from the upper and lower epidermis by parenchyma and collenchyma tissues. The vascular tissue of the small veins is surrounded by a parenchymatous sheaf; a few collenchyma cells sometimes occur between the sheaf and the lower epidermis.

In surface view, the upper and lower epidermal cells are irregular in shape and wavy-walled except near and over the veins where they are nearly rectangular, and over the midrib where they are oblong-rectangular and straight-walled. The walls of the cells over the midrib are faintly beaded. Stomata occur in both the upper and lower epidermis; the guard cells are surrounded by 2 epidermal cells.

Nonglandular, uni- and multicellular, papillose, pointed hairs occur on both sides of the leaf but they are more numerous on the upper side. They arise from the epidermis of the midrib, large veins, and leaf margins.

The capitate hairs have a very short stalk, usually less than 10 microns, and a bicellular, biseriate head about 35 microns in diameter. The oil glands have 4 cells, and are about 80 microns in diameter. They occur on both the upper and lower epidermis but are more numerous on the lower.

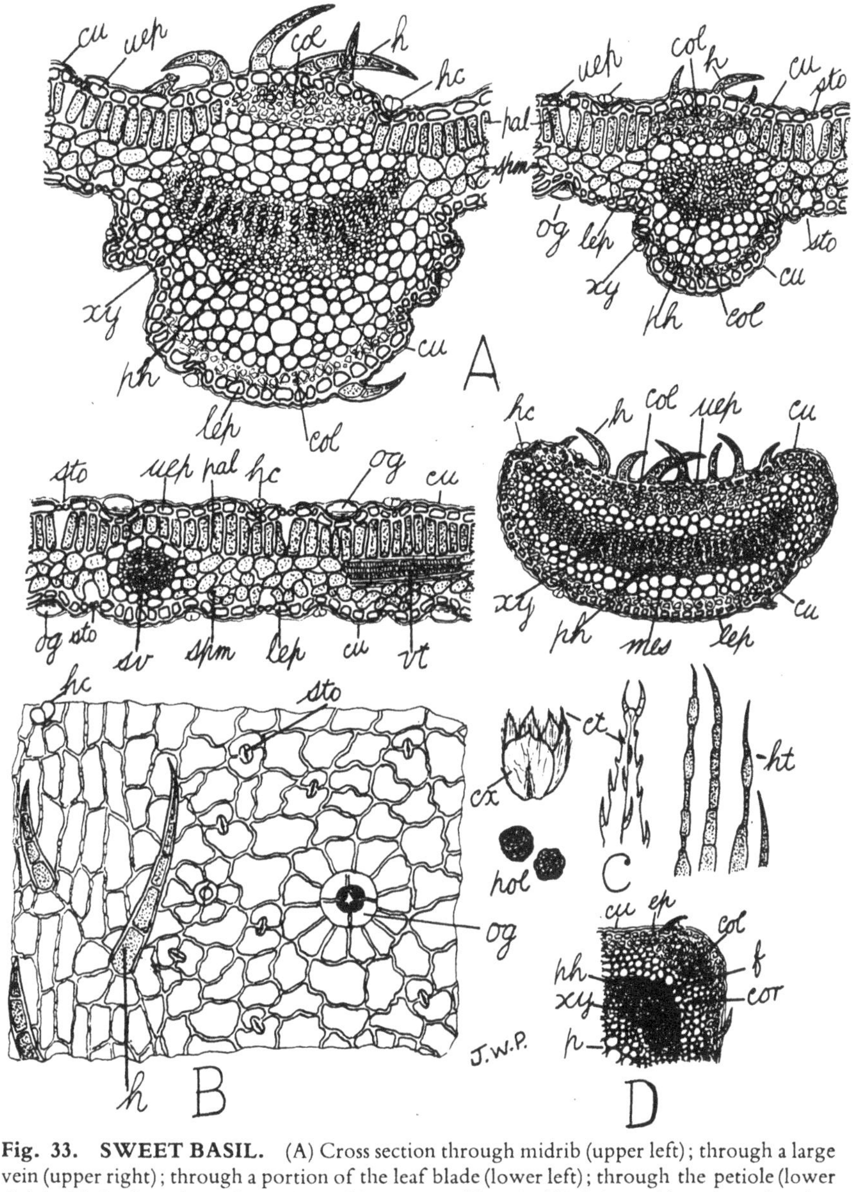

Fig. 33. SWEET BASIL. (A) Cross section through midrib (upper left); through a large vein (upper right); through a portion of the leaf blade (lower left); through the petiole (lower right). (B) Surface view of a portion of the upper epidermis. (C) Calyx with a view of a calyx tooth, throat hairs, and pollen. (D) Cross section through a corner of stem.

col, collenchyma; cor, cortex; ct, calyx teeth; cu, cuticle; cx, calyx; ep, epidermis; f, fibers; h, hair; hc, capitate hair; ht, throat hairs; lep, lower epidermis; mes, mesophyll; og, oil gland; p, pith; pal, palisade cells; ph, phloem; pol, pollen; spm, spongy mesophyll; sto, stoma; sv, small vein; uep, upper epidermis; vt, vascular tissue; xy, xylem.

Petiole

Cross section. The petiole is somewhat crescent-shaped. Nonglandular and glandular hairs are present. A layer of collenchyma tissue underlies the epidermal cells. The cells of the mesophyll are isodiametric and up to about 45 microns. A crescent-shaped group of xylem and phloem cells occupies a central position in the mesophyll.

Flowering Tops

Calyx. The calyx is gamosepalous with a broadly cordate, strongly veined, labiate lobe, very hairy on the surface toward the calyx tube. The calyx teeth are prominent and give rise to stiff, papillose hairs, uni- or multicellular with the latter predominating. Nonglandular hairs, capitate hairs, and oil glands occur on the outer epidermis. Capitate hairs and oil glands occur on the inner epidermis, and numerous long, multicellular, faintly papillose hairs arise from the throat of the calyx. In surface view, the cells of the inner and outer epidermis are small and wavy-walled; stomata are present.

Corolla. The corolla is gamopetalous. In surface view, the epidermal cells toward and at the base of the corolla tube are rectangular and straight-walled; toward the lobes they are oblong-rectangular and wavy-walled, and those of the lobes are small, irregular in shape and wavy-walled. Small, prismatic crystals occur in the lobe cells. Oil glands and numerous multicellular, faintly papillose hairs are present on the outer epidermis of the corolla.

Stamens and Style. The stamens are 4 in number and didynamous. The epidermal cells of the anthers are papillose and contain rosette crystals of calcium oxalate. The pollen grains are spherical with 6 divisions; they vary in size from about 45 to 55 microns. The style has a bilobed, papillose stigma.

Stem

Cross section. The stem is square or nearly so. The epidermal cells are rectangular. Hairs and oil glands are present. Collenchyma cells occupy the corners, and a layer of collenchyma tissue underlies the epidermis on all sides of the stem. The cortex cells are isodiametric and up to about 66 microns. A broken ring of fibrovascular tissue surrounds the pith. The pith consists of thin-walled, isodiametric cells up to about 88 microns.

XXXVI. THYME

Thymus vulgaris L.
Family: *Labiatae*

MORPHOLOGY

Dried leaves, greenish-brown, grayish on lower surface; small, reaching about 9 or 10 mm at most; ovate lanceolate, margin revolute, apex acute, petiolate; net venation, but obscure; glandular punctate; pubescent. The dot-like oil glands and hairs visible with a hand lens.

Floral parts: Calyx small, about 3 or 4 mm long, gamosepalous, upper lobe with 3 short teeth, lower lobe with 2 long, awl-like teeth deeply cleft; calyx throat hairy. Corolla frequently present, gamopetalous, 5 or 6 mm long, bilabiate; stamens 4, didynamous; style with bilobed stigma.

Odor fragrant and aromatic, accentuated when the leaves are crushed; taste aromatic, warm, and pungent.

HISTOLOGY

Leaf

Cross section. A cuticle occurs on both surfaces. The cells of the upper and lower epidermis are rectangular or nearly so; some pitting occurs in the cells of the upper epidermis. Nonglandular and glandular hairs are present. The nonglandular hairs are uni- or multicellular (most consist of 1 or 2 cells, but 3-cell hairs occur occasionally), thick-walled, strongly papillose with the exception of the basal cell of the multicellular hairs, some curved but mostly straight and stiff, pointed, clear or with dark contents. The capitate hairs have either 1- or 2-celled stalks. The mesophyll consists of 1 to 2 layers of palisade cells which vary in size, some reaching 90 microns, long axis, and spongy tissue with large intercellular spaces. Fibrovascular tissue traverses the spongy mesophyll. Numerous fibers are associated with the xylem and phloem of the midrib and veins. There is no collenchyma tissue.

In surface view, the epidermal cells of both sides of the leaf are irregular in shape with wavy walls except over the midrib and veins where the cells are rectangular and the walls straight. Stomata are few in the upper epidermis, abundant in the lower; the guard cells are surrounded by 2, 3, or 4 epidermal cells, with the 2-cell arrangement predominant. The capitate hairs are small, up to about 18 microns diameter. The oil glands are large, about 65 microns diameter, and 12-celled.

Petiole

Cross section. The epidermal cells are rectangular or nearly so. Hairs and oil glands are present. Collenchyma tissue occupies the corners and also forms a layer under the lower epidermis. Collenchyma tissue occurs between the fibrovascular bundle and the upper epidermis, but does not form a continuous layer under the upper epidermis. The fibrovascular tissue occupies a central position in the mesophyll.

Flowering Tops

Calyx. The calyx is gamosepalous, 5-toothed, and small. Two of the teeth are deeply cleft, project in prong-like fashion, and are prominent. Nonglandular and glandular hairs similar to those of the leaf occur on the outer epidermis. Numerous, long, multicellular, papillose hairs arise from the calyx throat. The inner epidermal cells in the region of the throat hairs are finely papillose.

Corolla. The corolla is gamopetalous. The apex epidermal cells of the lobes are papillose. Nonglandular and glandular hairs occur on the outer epidermis.

Stamens and Style. The stamens are 4 in number and didynamous. The epidermal cells of the anthers are papillose. The pollen grains are granular, have 6 furrows, and are spheroidal with diameter from 28 to 35 microns. The style has a bilobed stigma with papillose epidermal cells.

Stem

Cross section. The stem is square or nearly so. The epidermal cells are rectangular or nearly so. Nonglandular and glandular hairs are present. Collenchyma tissue occupies the corners and underlies the epidermis on all sides of the stem. The cortex cells are irregular in shape and thin-walled. The fibrovascular tissue forms an unbroken ring around the pith. The cells of the pith are thick-walled and pitted.

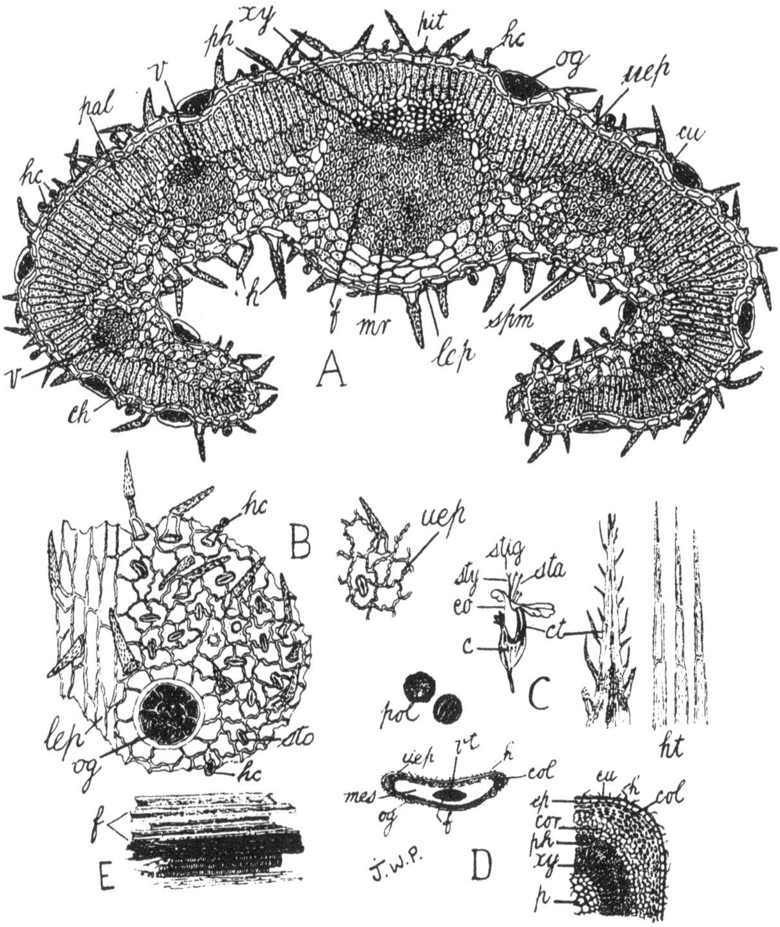

Fig. 34. THYME. (A) Cross section of leaf blade. (B) Surface view of lower epidermis (left); of upper epidermis (right). (C) Flower (left); calyx tooth and throat hairs (right). (D) Cross section of petiole (above); of corner of stem (right). (E) Fibrovascular tissue.

c, calyx; ch, chloroplasts; co, corolla; col, collenchyma; cor, cortex; ct, calyx tooth; cu, cuticle; ep, epidermis; f, fibers; h, hair; hc, capitate hair; ht, throat hairs; lep, lower epidermis; mes, mesophyll; mr, midrib; og, oil gland; pal, palisade cells; ph, phloem; p, pith; pit, pitted; pol, pollen; spm, spongy mesophyll; sta, stamen; stig, stigma; sto, stoma; sty, style; uep, upper epidermis; v, vein; vt, vascular tissue; xy, xylem.

XXXVII. LAUREL LEAVES

Laurus nobilis L.
Family: *Lauraceae*

[ALSO CALLED BAY LEAVES]

GENERAL

An evergreen tree. Leaves alternate, petiolate, firm; flowers yellow; fruit a berry.

MORPHOLOGY

The dried leaves are greenish-brown to light brown on the upper surface, lighter on the lower surface; Mediterranean leaves from about 2.5 to 7.5 cm in length, Guatemala leaves to about 12 cm. Leaves elliptical, acute to moderately acuminate at the apex, acute at the base, entire or undulating, stiff in texture, brittle, glabrous; upper surface smooth, lustrous; lower surface dull; net venation, pinnate; midrib prominent; petiole short.

Odor pleasant; fragrant and delicately aromatic when the dried leaves are crushed. Taste, aromatic and bitter.

HISTOLOGY

Leaf

Cross section. A moderately thick cuticle covers the epidermis. The cells of the upper epidermis are rectangular in shape or nearly so, with thick outer walls. The cells of the lower epidermis are similar to those of the upper but

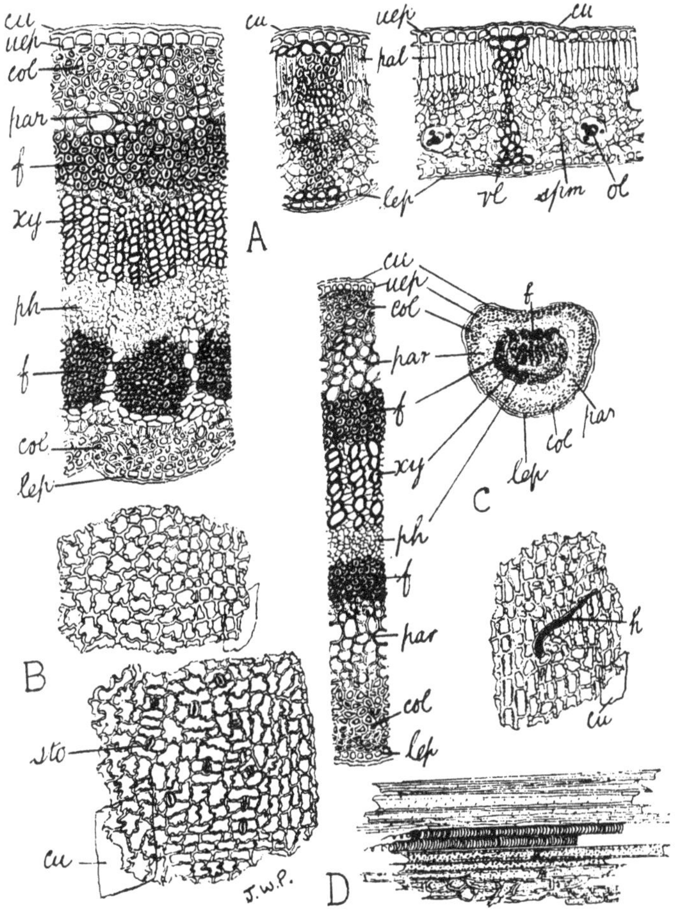

Fig. 35. LAUREL (BAY) LEAF. (A) Cross section through midrib (left); through vein (top, center); through portion of leaf blade (top, right). (B) Surface view of upper epidermis (above); of lower epidermis (below). (C) Cross section of petiole (above and left); surface view of epidermis of petiole (below). (D) Fibrovascular tissue.

col, collenchyma; cu, cuticle; f, fibers; h, hair; lep, lower epidermis; ol, oil; pal, palisade cells; par, parenchyma cells; ph, phloem; spm, spongy mesophyll; sto, stoma; uep, upper epidermis; vl, veinlet; xy, xylem.

the outer walls are thinner. The mesophyll consists of 2 layers of palisade cells, spongy tissue, oil cells, fibrovascular and collenchyma tissue. The palisade cells vary in size from about 33 to 44 microns, long axis. The oil cells occur in the spongy mesophyll. The fibrovascular bundle of the midrib is large, with the vessels arranged in radial rows. Numerous fibers accompany the phloem and xylem tissues. Collenchyma tissue occurs above and below the fibrovascular bundle of the midrib but does not accompany the vascular tissue of the veins.

In surface view, the cells of the upper epidermis are irregular to oblong-rectangular in shape, always the latter near and over the midrib and veins. The cell walls are sinuous, thick, and porous. The cells of the lower epidermis are similar in shape to those of the upper but larger and with thinner walls. There are no stomata on the upper surface. Stomata are numerous on the lower surface; the guard cells are surrounded by 4 epidermal cells.

Petiole

Cross section. A moderately thick cuticle covers the epidermis. The epidermal cells are rectangular in shape; the outer walls are thick. Occasional unicellular hairs are present. The mesophyll consists of parenchyma cells with moderately thick walls, and oil cells. The fibrovascular tissue occupies a central position in the mesophyll.

In surface view, the cells of the epidermis are irregular to oblong-rectangular in shape; the walls are thick. Occasional, curved, pointed, unicellular hairs up to about 220 microns in length and with blackish contents occur on the epidermis.

XXXVIII. PARSLEY

Petroselinum sativum Hoffm.
Family: *Umbelliferae*

GENERAL

This is a biennial, herbaceous plant with rich green leaves and small white flowers; fruit, a schizocarp.

MORPHOLOGY

Leaves pinnately compound, leaflets 3-lobed, each lobe divided into smaller, tooth-like lobes or segments; curly; green. The commercial product consists of the dried leaves, broken, shrivelled, curled.

Dried parsley leaves have a characteristic, agreeable, aromatic odor and taste.

HISTOLOGY

Leaflet

Cross section. A thin, striated cuticle covers the epidermis. The cells of the upper epidermis are nearly rectangular in shape, and vary in size up to about 44 microns, long axis, except over the main veins where they are larger and frequently ballooned. The cells of the lower epidermis are similar to those of the upper but less rectangular and not ballooned over the veins. The mesophyll consists of a layer of palisade cells, and spongy tissue. The palisade cells are around 40 to 55 microns, long axis, in the mature leaflet. Vascular bundles traverse the spongy mesophyll. Collenchyma tissue is present over and under the vascular bundles but is more abundant below, i.e., toward the lower epidermis. Adjacent to the phloem tissue there is a vitta or oil cavity, characteristic of the *Umbelliferae*.

In surface view, except over the veins the cells of the upper and lower epidermis are irregular in shape with wavy walls, and vary in size to about 70 microns, long axis, with the occasional elongated cell exceeding this length. Over the veins the upper epidermal cells are mostly ballooned; the lower epidermal cells are oblong-rectangular or elongated and tapered in shape, and vary in size to about 90 microns, long axis. Stomata are few in the upper epidermis, numerous in the lower; the guard cells are surrounded by 2, 3, or 4 epidermal cells.

Rachis

Cross section. The rachis, from which the leaflets arise, consists of (a) cuticle; (b) an upper and lower epidermis; (c) a large central fibrovascular bundle with thin-walled fibers. In longitudinal view the fibers are long with thin walls; the vessels are spiral; (d) outer and inner mesophyll tissue of small parenchyma cells with chloroplasts; (e) central mesophyll tissue of large, empty parenchyma cells; (f) oil cavities or canals; and (g) collenchyma tissue. The rachis is ribbed, and the collenchyma cells occur in groups at each rib; they give strength and rigidity to the rachis. The collenchyma is laid down thickly.

In surface view, between the ribs the cells of the epidermis are elongated, and the cell walls curve in a slightly wavy or sinuous way. Stomata are present and are surrounded by 2, 3, or 4 epidermal cells with the 3-cell arrangement predominant. Over the ribs and collenchyma tissue the epidermal cells are oblong-rectangular in shape or nearly so. They vary in size, with some reaching 200 microns, long axis. The cell walls are straight, beaded in places, and thicker than those of the cells between the ribs. All of the epidermal cells are striated and covered with a striated cuticle.

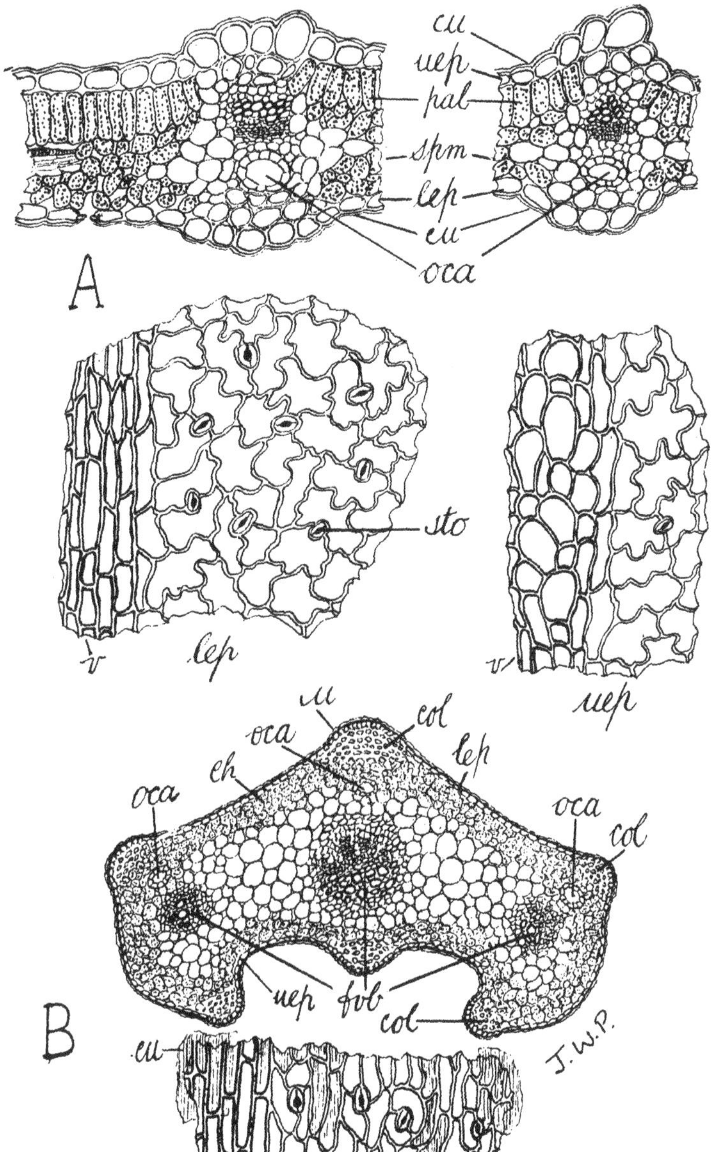

Fig. 36. PARSLEY. (A) Cross section through a large vein of leaflet (above, left); and through a small vein (above, right). Lower epidermis between veins, and over a vein, surface view (below, left); upper epidermis between veins, and over a vein, surface view (below, right). (B) Cross section of rachis (above); epidermis of rachis, surface view (below).

ch, chloroplast cells; col, collenchyma; cu, cuticle; fvb, fibrovascular bundle; lep, lower epidermis; oca, oil cavities; pal, palisade cells; spm, spongy mesophyll; sto, stoma; uep, upper epidermis; v, vein zone.

XXXIX. CHERVIL

Anthriscus cerefolium Hoffm.
Family: *Umbelliferae*

GENERAL

Chervil is an annual herb, propagated by seed. The seed is sown in the early spring and the plants are ready for harvesting in April or May. The plant ordinarily reaches a height of 45 to 60 cm, but as cultivated for commerce in the United States the plants range from 25 to 30 cm in height at maturity. Chervil has decompound leaves and small white flowers arranged in compound umbels of 3 to 5 slender rays. The lateral umbels are sessile and the terminal umbels pedunculated. The involucre is composed of 1 to 3 bracts, the involucels of 2 to 3 bracts. The involucre is sometimes absent.

Chervil is cultivated for the spice trade in France and in the United States. In France the leaves are dehydrated by circulating air warmed to a temperature of 50° C. The leaves are fragmented by mechanical means and packed in plastic lined containers. The net weight of the package is 12½ kilograms (27½ lbs). The French product is exported to all parts of Europe. In the United States the leaves are similarly processed for the American spice trade. There are no flowering tops in the commercial product.

MORPHOLOGY

Leaves decompound with more or less oval, cut leaflets; leaves alternate, tender, thin, pale green; 2 to 4 cm long and 2 to 4 cm wide; net venation, pinnate; a few hairs on the midvein and margin of leaflets, visible only with the microscope. As cultivated for commerce in the United States the leaves are about 1.6 to 2.5 cm in length and 1.6 to 2 cm in width.

The odor and taste of dried chervil is faintly reminiscent of a blend of caraway and anise.

HISTOLOGY

Cross section of leaflet: A thin cuticle covers the upper and lower epidermis; it is striated at the margin and over the lower epidermis of the midvein. The cells of the upper and lower epidermis are transversely irregularly oblong-rectangular, or nearly so, and some cells of the midvein have a slight tendency to balloon, more so in the lower epidermis. The mesophyll consists of a single row of palisade cells in the upper part and a spongy mesophyll of parenchyma cells in the lower. The palisade cells vary in size from about 18 to 75 microns, long axis, and from about 16 to 35 microns, short axis. The cells of the spongy mesophyll vary in shape; many are elliptical, or nearly so, some have a tendency to transverse oblong-elongation, some are isodiametric, and others are irregular. The palisade and spongy mesophyll cells contain chloroplasts. The vascular system traverses the spongy mesophyll. Oil ducts or secretory canals, circular in shape, accompany the main veins and are situated adjacent to and below the phloem tissue of the vascular bundle.

In surface view, except over the margin and main veins, the cells of the upper and lower epidermis are irregular in shape, and they vary in size to about 110 microns, long axis. The cell walls are undulating, those of the lower epidermis being the more wavy. The epidermal cells of the main veins and margin, particularly those of the lower epidermis of the midvein, are longitudinally oblong, oblong-pointed, or irregularly angled, with nearly straight or slightly curved walls. The walls of the epidermal cells of the midvein are faintly beaded, more distinctly so toward the base of the leaflet. Stomata are numerous in the lower epidermis and almost absent in the upper, an examination of various areas of the upper epidermis of a number of leaflets showing no stomata or only 1, 2, or 3. In size the greater number of stomata are about 23 microns, long axis, and the guard cells are surrounded by 2, 3, 4 or 5 epidermal cells. Hairs arise from the epidermal cells of the main veins and the margin and occur on both the upper and lower epidermis but are more numerous on the lower; they are not abundant. The hairs on the upper epidermis are very few in number and occur for the most part on the margin. Except for a very occasional bicellular type the hairs are unicellular, clear, and pointed. In size they vary in length to about 550 microns and are about 45 microns broad at the base. Short, sharp, tooth-like hairs emerge at irregular intervals from the edges of the lobes of the leaflet, and most lobes terminate in such an emergence.

Rachis

Cross section of rachis: A thin, striated cuticle covers the epidermis. The cells of the upper and lower epidermis are irregularly oblong-rectangular, or nearly so, and occasionally almost isodiametric in shape. They vary in size from about 10 to 23 microns, long axis, with outer walls from about 5 to 7 microns in thickness. The mesophyll consists of parenchyma cells with small intercellular spaces. A few collenchyma cells occur immediately within the epidermis over the ribs. There are three vascular bundles, the largest being centrally situated in the mesophyll; a small bundle is situated in the region of each upper corner. An oil duct accompanies each vascular bundle. The oil duct accompanying the large central bundle is situated adjacent to and below the phloem tissue. The ducts acompanying the upper bundles are situated adjacent to and at the outer side of the phloem tissue.

In surface view, the epidermal cells are for the most part longitudinally oblong-rectangular or oblong-pointed in shape with nearly straight or slightly curved walls.

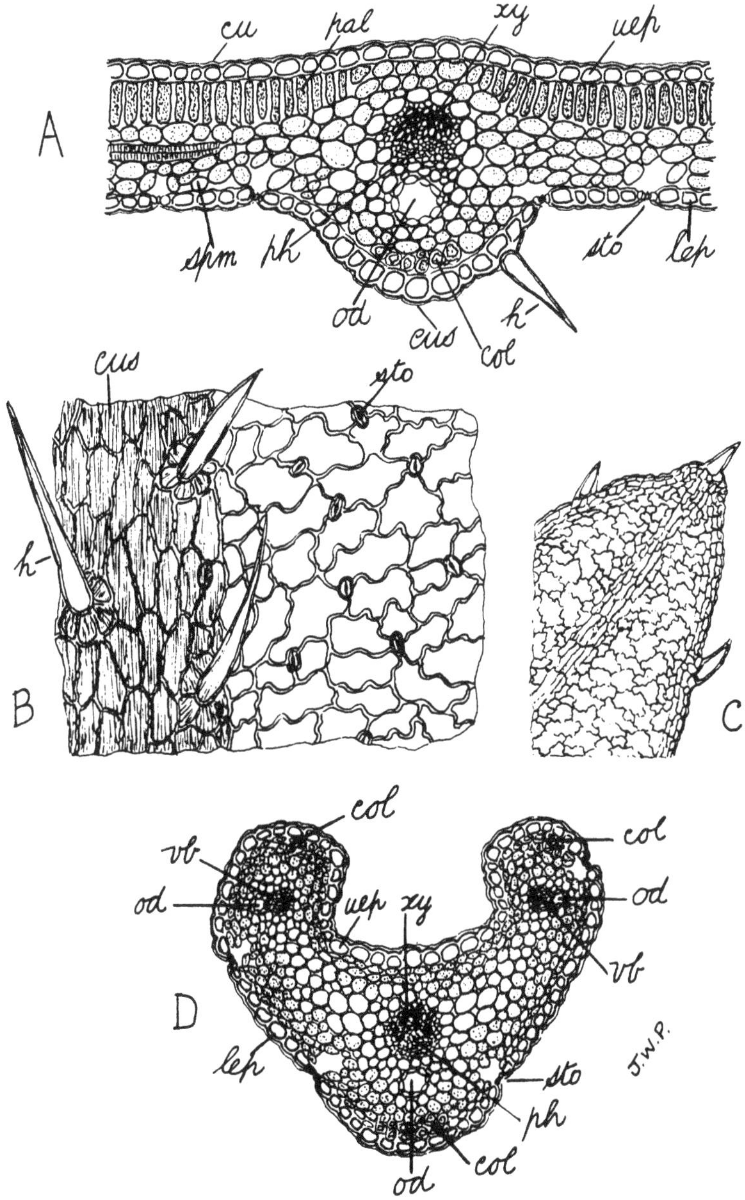

Fig. 37. CHERVIL. (A) Cross section of a main vein of a leaflet. (B) Lower epidermis, surface view. (C) Apex of a lobe of a leaflet. (D) Cross section of rachis.

col, collenchyma; cu, cuticle; cus, striated cuticle; h, hair; lep, lower epidermis; od, oil duct; pal, palisade cells; ph, phloem; spm, spongy mesophyll; sto, stoma; uep, upper epidermis; vb, vascular bundle (vein); xy, xylem.

XL. TARRAGON

Artemisia dracunulus L.
Family: *Compositae*

GENERAL

There are two varieties of tarragon known to commerce—namely, Russian tarragon and French tarragon. The former is reported to have coarser leaves than the latter, often called the true tarragon, and a less desirable flavor.

Tarragon produced in France enjoys wide distribution in Europe.

French tarragon reaches a height of about 70 cm and has small, very pale yellow, generally sterile flowers. Because of the latter, propagation is accomplished by root division or by cuttings. At maturity, the leaves are harvested, dried by circulated warm air and brought to the desired degree of fragmentation by mechanical grinding. The prepared dried herb is packaged in plastic bags and kraft paper for delivery in France and for export to other European countries. The package has a net weight of 12½ kilograms (27½ lbs).

It is the French tarragon variety that is cultivated in the United States for the North American spice trade. It is propagated as above, harvested at maturity, and dehydrated with controlled artificial heat and air to maintain both color and aroma. The dried leaves are mechanically fragmented with a series of machines which mill, brush, shake, blow and screen the leaf particles through specified apertures.

MORPHOLOGY

Leaves lanceolate, entire, sessile, and (with the exception of a few minute capitate hairs) glabrous; bright green; net venation, pinnate; midrib and margin ribs prominent. The edges of the dried leaves are curled toward the lower surface; finely revolute. In size, the whole dried leaves vary to about 4 cm in length, and in width to about 3.5 mm.

Odor and taste: Agreeable, aromatic and reminiscent of anise.

HISTOLOGY

Leaf

Cross section of the leaf: A thin cuticle, striated over the midrib and marginal veins, covers the upper and lower epidermis. The cells of the upper and lower epidermis are transversely oblong-rectangular, or nearly so. They vary in size from about 18 to 50 microns, long axis, and from about 10 to 30 microns, short axis. The outer walls of the midrib epidermal cells are moderately thick, being about 10 microns on the upper side and 11 or 12 microns on the lower. An occasional, short capitate hair, or the remains of one, may be seen. Stomata are present in both the upper and lower epidermis.

The mesophyll consists of two rows of palisade cells in the upper part and one row in the lower, and a central spongy mesophyll of parenchyma cells with intercellular spaces. The palisade cells vary in size from about 36 to 90 microns, long axis, and from about 18 to 22 microns, short axis. The palisade and spongy mesophyll cells contain chloroplasts. The vascular tissues traverse the spongy mesophyll. As in most leaf cross sections, veins cut transversely are seen in some areas of the spongy mesophyll. Sometimes, 3 or 4 fibers accompany the phloem of the midrib.

A few collenchyma cells usually occur immediately within the lower epidermis of the midrib of mature leaves. Oil ducts or secretory canals, more or less circular in shape, occur adjacent to the vascular bundles. In some cross sections as many as 7 or 8 ducts are seen, in others 5 or 6. Usually, there are two ducts in the midrib area, one of which is situated above and to the left of the vascular bundle, the other above and to the right of the bundle; one duct situated between the two small vascular bundles of the leaf margin, and one duct accompanying each of the veins between the midrib and margin. However, in some sections there may be only one vascular bundle in the margin with an oil duct situated at the side or above the bundle. There is not an absolute uniformity or regularity of arrangement.

In surface view, except over the midrib and margin veins, the cells of both the upper and lower epidermis are irregular in shape and vary in size to about 113 microns, long axis. The cell walls are undulating, those of the lower epidermis being the more wavy. The epidermal cells of the midrib and margin veins are longitudinally oblong-rectangular, oblong pointed, or angular with nearly straight, beaded walls. They vary in size to about 145 microns, long axis. Stomata are numerous in both upper and lower epidermis with the greater number being in the lower, about 25% more. They are about 36 microns, in size, long axis, and the guard cells are surrounded by 3, 4, and occasionally 5, epidermal cells. Occasional, small, inconspicuous capitate hairs emerge from the epidermis. They are short and about 35 to 40 microns in diameter. Very few perfect specimens of these hairs remain on the dried leaves of the commercial product.

Stem

Cross section of stem: A thin, striated cuticle covers the epidermis. Two rows of collenchyma cells underlie the epidermis and these increase in number to five or six rows in the corners of the stem. The remaining cortex cells are parenchymatous with intercellular spaces. These cells vary in size to about 36 microns, long axis. The cortex cells contain chloroplasts. An endodermis consisting of an unbroken sheath of cells envelops the vascular system and pith. In prepared sections, these cells show no contents. They are tangentially oblong, or nearly so, in shape, and vary in size from about 22 to 46 microns, long axis. Oil ducts or secretory canals occur in the concave areas of the parenchymatous cortex between the ribs and adjacent to the endodermis. They are more or less elliptical in shape, and vary in size from about 45 to 68 microns, long axis. Fibrovascular bundles grouped closely together encircle the pith. The pith consists of parenchyma cells with pitted walls. These cells vary in size to about 68 or 70 microns, long axis, and occasionally larger.

In surface view, the cells of the epidermis are longitudinally oblong, oblong-pointed, angular, or irregularly angled with nearly straight or curved, beaded walls. Stomata are present.

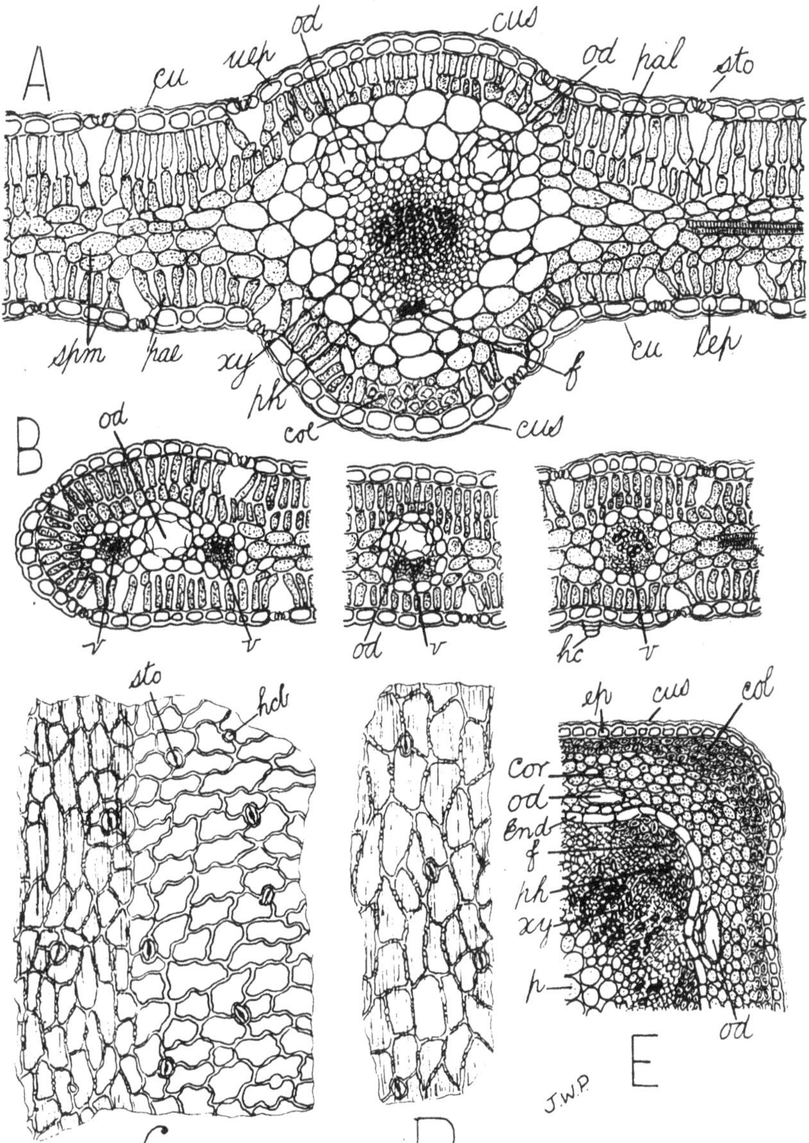

Fig. 38. TARRAGON (A) Cross section through midrib. (B) Cross section of leaf margin (left); portion of leaf blade through a vein with oil duct (center); and of a vein with no oil duct (right). (C) Upper epidermis, surface view. (D) Epidermis of stem, surface view. (E) Cross section of a portion of stem.

col, collenchyma; cor, cortex; cu, cuticle; cus, striated cuticle; end, endodermis; ep, epidermis; f, fibers; hc, capitate hair; heb, capitate hair base; lep, lower epidermis; od, oil duct; p, pith; pal, palisade cells; ph, phloem; spm, spongy mesophyll; sto, stoma; uep, upper epidermis; v, vein; xy, xylem.

Bulbs

XLI. GARLIC

Allium sativum L.
Family: *Liliaceae*

GENERAL

This is a hardy, perennial, bulbous plant with long, flat, linear leaves, and purple flowers. The bulblets are also known as "cloves" of garlic.

MORPHOLOGY

Dried, subglobular, compound bulb consisting of several bulblets contained within whitish, membranous scales attached to a short, disk-like, woody stem with numerous rootlets on the underside. Bulblets irregularly ovoid, dorsal surface convex, each consisting of 2 or 3 yellowish-green leaves folded longitudinally and contained within 2 white, fleshy, modified leaf bases or scales surrounded by 2 thin, whitish scales, the outermost being very thin, papery, and brittle.

The spice is the ground product of dehydrated garlic, and sold as Garlic powder. In the finer grades only the selected cloves of garlic are used. The powder is creamy-white or cream, and free running.

Odor strongly alliaceous; taste intensely pungent.

HISTOLOGY

Bulblet

Outer Scale. Cross section. The outer scale consists of (a) an outer epidermis; (b) an hypodermal crystal layer underlying the outer epidermis; (c) several layers of mesophyll parenchyma cells; (d) vascular tissue; and (e) an inner epidermis.

In surface view, the cells of the outer epidermis are elongated, narrow, with thin, porous walls; those of the inner epidermis are elongated, narrow, with thin walls. The cells of the hypodermal crystal layer are ellipsoidal with thick, porous walls; each cell contains a large, prismatic crystal of calcium oxalate, and many cells have, in addition, some small, prismatic crystals of the same substance. The long axes of the large crystals tend to lie parallel, or nearly so, to the long axes of their cells; and the long axes of the cells lie rectangularly to the long axes of the overlying epidermal cells; this arrangement holds true in general. The vessels are spiral. The veins run parallel to each other.

Inner Scale. The inner scale consists of the same number and type of tissues as the outer scale with following differences: The cells of the outer epidermis are markedly sclerenchymatous; in surface view, they are elongated, with greatly thickened walls and very narrow lumen. The cells of the hypodermal crystal layer, in surface view, are somewhat smaller than those of the outer scale, the walls are more frequently pitted, and the crystals are slightly smaller.

Fleshy Leaf Bases. Two whitish, fleshy, modified leaf bases form the bulk of the bulblet and enclose the small foliage leaves. The thickness of the outer leaf base is considerably greater than that of the inner. Both consist of an outer and inner epidermis, mesophyll, and scattered vascular bundles.

In surface view, the cells of the outer epidermis are thin-walled, tend somewhat to the oblong-rectangular in shape, and are about 30 microns in width, and up to about 90 microns in length. The cells of the inner epidermis are thin-walled, less rectangular than those of the outer epidermis, and are about 55 microns in width, and up to about 275 microns in length. The mesophyll consists of large, thin-walled parenchyma cells, mostly polygonal in shape, with small intercellular spaces. The vessels are spiral.

Foliage Leaves. There are usually 3 of these leaves developed. The upper and greater part of the leaf is greenish-yellow, the lower part is whitish. In cross section, the greenish-yellow portion of the leaf is complete with epidermal tissue including stomata; a mesophyll of isodiametric, chloroplast-containing cells underlying the outer and inner epidermis, with a central zone of polygonal cells without chloroplasts. Vascular tissue traverses the mesophyll.

Stem

The stem is the small, disk-like base of the bulblet. It emerges from the hard, woody stem of the compound bulb from which the thin, papery scales enveloping the bulblets arise, and on which the developed rootlets are seen. It consists of parenchyma cells, scattered vascular tissue, and the rudiments of adventitious roots.

If the woody, rootlet-bearing stem of the compound bulb is not removed and discarded in the preparation of garlic powder, the product will contain, in addition to the tissues described above, stone cells, sclerenchymatous pitted cells, many of the latter containing prismatic crystals of calcium oxalate, and spiral and reticulate vessels of the xylem tissue; all present in that part of the plant.

The developed rootlet consists of epidermis, exodermis, cortex, endodermis, pericycle, phloem and xylem tissues.

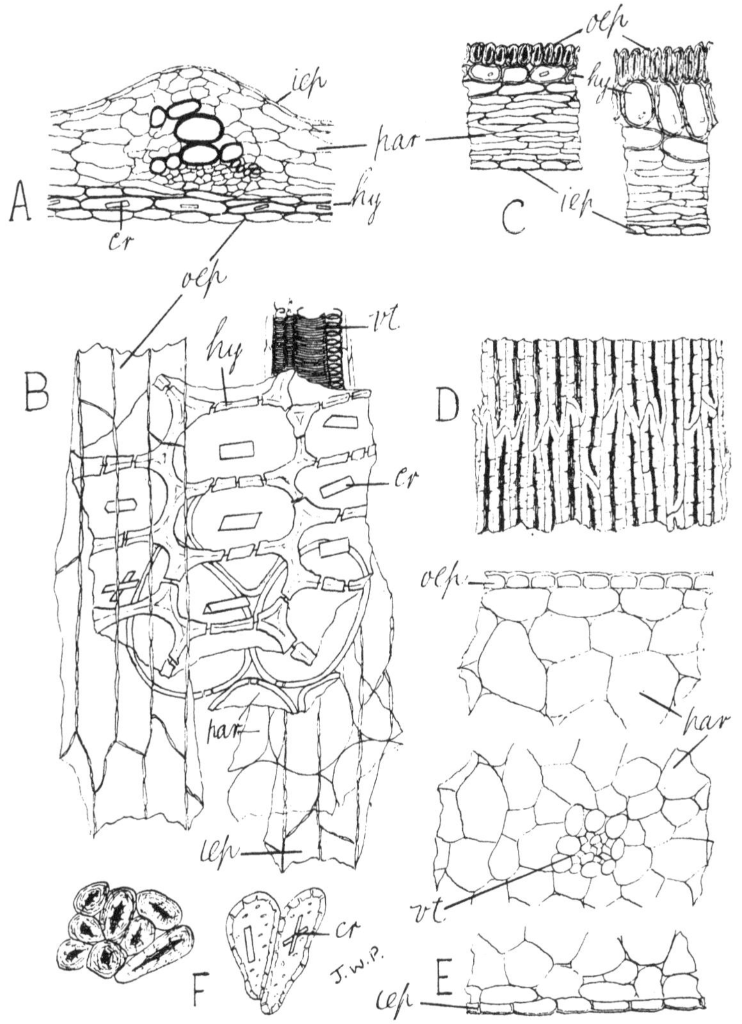

Fig. 39. GARLIC. (A) Outer leaf scale, cross section through vascular bundle. (B) Outer leaf scale, surface view of outer epidermis, hypodermal crystal layer, parenchyma, vascular tissue, and inner epidermis. (C) Cross section of a portion of inner leaf scale. (D) Surface view of sclerenchymatous outer epidermis of inner leaf scale. (E) Cross section of fleshy leaf base. (F) Stone cells, and sclerenchymatous pitted cells with crystals, from woody stem of compound bulb.

cr, crystal; hy, hypodermal layer; iep, inner epidermis; oep, outer epidermis; par, parenchyma; vt, vascular tissue.

XLII. ONION

Allium cepa L.
Family: *Liliaceae*

GENERAL

This is a biennial, bulbous plant with long, slender, dark green tubular leaves; flowers small, white or greenish-white, borne in a globular head on a tubular stalk.

MORPHOLOGY

A tunicate bulb consisting of several white, fleshy, modified leaf bases or scales surrounded by pigmented membranous scales. Stem small, disk-like, and bearing numerous rootlets on the underside.

The spice is the ground product of dehydrated onions, and sold as onion powder. It is creamy-white in color and free running.

Odor alliaceous; taste pungent, sharp.

HISTOLOGY

Membranous Leaf Scales

Cross section. This consists of (a) an outer epidermis; (b) a hypodermal crystal layer underlying the outer epidermis; (c) several layers of collapsed mesophyll parenchyma cells; (d) vascular tissue; and (e) an inner epidermis.

In surface view, the epidermal cells are elongated and narrow. The walls of the epidermal cells are thin; those of the outer epidermis are slightly thicker than those of the inner. The cells of the hypodermal crystal layer are large and regularly arranged; the walls are thicker than those of the overlying epidermal cells. Each cell contains one large, prismatic crystal of calcium oxal-

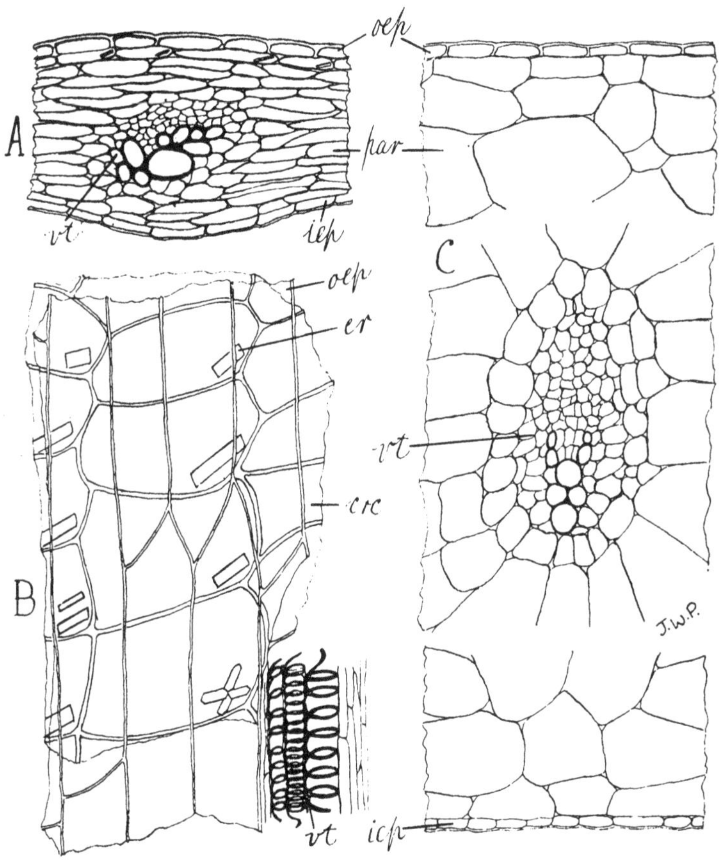

Fig. 40. ONION. (A) Leaf scale, cross section through vascular bundle. (B) Leaf scale, surface view of outer epidermis, crystal layer, and vascular tissue. (C) Fleshy leaf base, cross section through a vascular bundle.

cr, crystal; crc, crystal cells; iep, inner epidermis; oep, outer epidermis; par, parenchyma tissue; vt, vascular tissue.

ate, and some cells have, in addition, one or more small, prismatic crystals of the same substance. The crystals usually occur in similar corners of the cells in which they are contained, with their long axes tending to lie diagonally to the long axes of the overlying epidermal cells. The long axes of the crystal cells lie rectangularly to the long axes of the epidermal cells. Vascular tissue traverses the mesophyll; the vessels are spiral.

Fleshy Leaf Bases

The whitish, modified leaf bases form the bulk of the bulb, and consist of thin-walled epidermal cells, elongated in surface view; thin-walled mesophyll parenchyma cells; and vascular tissue with spiral vessels.

Stem

The shallow, disk-like stem situated at the base of the bulb consists of parenchyma cells, vascular tissue, and the rudiments of adventitious roots. Spiral and reticulate vessels occur in the vascular tissue.

Part 2

CHEMICAL COMPOSITION

INTRODUCTION

Spices are dried plant products which add flavor, relish, or piquancy to foods. They consist of rhizomes, barks, leaves, fruits, seeds, and other parts of plants. Most are fragrant, aromatic, and pungent.

The bulk of the dry matter of these products consists of carbohydrates—organic compounds of carbon, hydrogen and oxygen, the hydrogen and oxygen usually in the same proportion as in water. Cellulose, starch, sugars, pentosans, mucilage are carbohydrates.

Cellulose is a carbohydrate used by the plant in the building of cell walls. Parenchyma and collenchyma tissues are cellulose, sclerenchyma tissue is a combination of cellulose and lignin, lignocellulose. The nature of lignin is not fully known. Cellulose is a complex organic compound, a polysaccharide composed of a long chain of monosaccharides, insoluble in water and in most solvents, hydrolyzed by sulfuric acid into a colloidal substance which gives a blue reaction with iodine, dissolved by a concentrated solution of zinc chloride, and by cuprammonia solution. It is tasteless.

Starch is a polysaccharide with the same basic formula as cellulose but less complex. Its starting point is glucose and it originates in leucoplasts (colorless plastids), the starch being laid down in layers around a central core or point of origin called the hilum. The granules have a characteristic size and shape depending upon the species of the plant, and, because of this, starch serves as a criterion for the detection of adulteration. Starch occurs in many of the spices, being abundant in ginger, nutmeg and pepper. A transitional starch product called amylodextrin occurs in mace. Starch reacts with a blue color, and amylodextrin with a reddish-brown color, when treated with iodine-potassium iodide solution on the slide. When boiled with water, the starch granules swell forming a solution colloidal in nature which when cool reacts with a dark blue color on the addition of a few drops of iodine-potassium iodide solution. Nutmeg (starch content) reacts as above when boiled with water; mace (amylodextrin content) when boiled with water reacts with a pale pinkish-brown color on the addition of iodine-potassium solution on cooling down.

Of the sugars, glucose and fructose are the most commonly found. Glucose probably occurs in all plants. It is a monosaccharide, hexose, readily soluble in water, and with a sweet taste. It undoubtedly enters into numerous complex chemical processes within the plant. Cellulose and starch yield glucose on complete hydrolysis. Glucosides, consisting of glucose combined with one or more organic substances are of common occurrence in plants. The decomposition of a glucoside within the plant, effected by an enzyme, may yield a substance of characteristic odor or taste. For example, the pungency of black mustard seed is due to allyl isothiocyanate, produced with glucose and potassium hydrogen sulfate when the glucoside sinigrin is decomposed by the action of the enzyme myrosin.

Pentosans are widely distributed in plants, and occur in many spices. They are polysaccharides closely related to cellulose, and consist of araban and xylan. On hydrolysis they yield the pentoses arabinose and xylose. Xylan occurs in lignified cell walls, in seed coats, and probably in other zones of lignified tissue. It is probably associated with structural, supporting, and protective processes.

Mucilage is a carbohydate whose chemistry is imperfectly known; it yields pentoses and hexoses when broken down. Layers of mucilage constitute a part of certain cell walls, hard and horny in the dry condition, but which form a thick, viscid, adhesive liquid when soaked in water. Mucilage occurs in mustard seeds, fenugreek, and some other spices.

Fixed oil constitutes a large part of some spice products, poppy and sesame seed, for example. It is present in endosperm and embryo tissue in a number of others. Like carbohydrates, the fixed oils contain carbon, hydrogen and oxygen, but have relatively less oxygen and more hydrogen. They are lighter than water, and leave a permanent, translucent stain on paper. Fixed oils dissolve readily in ether and chloroform. In general, they have no particular odor or taste. Fixed oils are compounds of glycerine with various fatty acids; esters of the trihydric alcohol, glycerol. Except those of mustard, poppy and sesame seeds, the fixed oils of the spices have no commercial value.

Proteins are complex organic compounds made up of simpler compounds known as amino acids. They consist of carbon, hydrogen, oxygen, and nitrogen; many contain sulfur, some contain phosphorous. In the storage organs of plants proteins commonly exist as aleurone grains; they are common in seeds and occur in all of the seeds of the Umbelliferae covered in this work, in fenugreek, mustard, poppy and sesame seeds, and in the seed of star anise; a protein substance occurs in the endosperm, and aleurone grains in the embryo, of cardamom seed.

Tannins are amorphous, colloidal, astringent substances. Their chemical constitution is not fully known. They combine with gelatine to form insoluble compounds; with ferric chloride solution they give a blue or green color. Tannins occur in many spices.

Resins are amorphous, solid or semisolid, complex substances. Associated with volatile oils they are known as oleoresins. Resins are insoluble in water, soluble in alcohol, ether, chloroform, and fixed oils. Resin occurs in many spices. The pungent principles of ginger and pepper are contained in their oleoresins.

Pigments are chemical compounds imparting color to the plant. The green color is caused by chlorophyll contained in chloroplasts within the plant cells. The reds, blues and purples are due to the presence of chemical compounds called anthocyanins. Yellow pigments, the carotinoids, are associated with chlorophyll in the chloroplasts, and also occur in chromoplasts. Pigments occur in most spices. The coloring matter of capsicum, turmeric, saffron and safflower are products of diverse and complex chemical compounds.

Calcium oxalate is widespread in plant organisms and occurs in many spice products. Prismatic crystals, large and small, and rosette crystals of calcium oxalate are very common. Very minute crystals deposited in powder-like form are known as crystal sand. The prismatic forms consist largely of needle-like, cubical, monoclinic, or tetragonal crystals, some very small, others very large.

Mineral elements. A formidable list of mineral elements occur in plants, but only a few are common, namely, potassium, sodium, calcium, magnesium, iron, phosphorous, sulfur, silicon, manganese, and chlorine. Mineral elements occur in all spices and may be determined by analysis of the ash. The ash is obtained by incinerating the spice.

The volatile or essential oils are important products of spices. They contain terpenes, sesquiterpenes, alcohols, esters, aldehydes, ketones, phenols, ethers, etc., and in some oils sulfur and nitrogen. Most volatile oils are slightly soluble in water, imparting to the water their odor and taste. They are miscible with absolute alcohol, soluble in various dilutions of alcohol depending upon the kind of oil, and soluble in ether and in chloroform. They evaporate cleanly from paper. The volatile oils occur in secretory cells, vittae, glands, and other special tissues, and are responsible for the aroma and taste of most spices. The grinding of spices breaks down or exposes these tissues with the inevitable loss of aromatic principles by evaporation of the volatile oil. For this reason ground spices should be freshly prepared, kept well covered, and if for retail sale packed into containers promptly.

The chemistry of the volatile oils has been a subject of study for many years, and such knowledge as we have of the chemical compounds contained in the volatile oils of the spices comes to us from the work of a large number of able investigators.

Following the text of Part 3 a short bibliography is given. To those who seek further information than this brief work presents on the chemical composition of the spices these books are recommended.

XLIII. THE SPICES

Allspice

Allspice contains up to about 4.5% of volatile oil, a fixed oil, tannin, resin, proteins, cellulose, pentosans, starch, pigments, mineral elements, etc.

United States Government Standards require that allspice shall contain "not less than 8% of quercitannic acid (calculated from the total oxygen absorbed by the aqueous extract) not more than 25% of crude fiber, not more than 6% of total ash, nor more than 0.4% of ash insoluble in hydrochloric acid."

Canadian Government Standard,* B.07.006: "ALLSPICE or PIMENTO, whole or ground, shall be the dried, nearly ripe fruit of the pimento tree, shall contain not less than 8% quercitannic acid, calculated from the total oxygen absorbed by the aqueous extract, and shall contain not more than

(a) 25% crude fiber.

(b) 6% total ash.

(c) 0.4% ash insoluble in hydrochloric acid."

The essential oil, known as Oil of Pimenta, is a colorless, yellow, or reddish-yellow liquid with the odor and taste of allspice; it becomes darker with age. Oil of Pimenta is very slightly soluble in water, soluble in glacial acetic acid, soluble in 1 volume of 90% alcohol, and soluble in 2 volumes of 70% alcohol. The chief constituent of the oil is eugenol, empirical formula $C_{10}H_{12}O_2$; structural formula $CH_2{:}CHCH_2C_6H_5(OCH_3)OH$; a liquid with an odor of cloves, and a pungent taste. Eugenol constitutes from about 60 to 75% of the volatile oil of allspice. Other organic reported compounds present in the oil include eugenol methyl ether, empirical formula $C_{11}H_{14}O_2$; cineole, $C_{10}H_{18}O$; *l*-α-phellandrene, $C_{10}H_{16}$; and caryophyllene, $C_{15}H_{24}$.

The quantity of volatile oil and other ether-soluble constituents present in allspice depends upon the quality and age of the product. Using the continuous extraction method the author determined the ether extract in three

*Division 7, Food and Drug Regulations, Food and Drug Directorate, Department of National Health and Welfare, Ottawa, Canada [3-9-1964].

samples of allspice as follows:

Extract	*Sample (a)* %	*Sample (b)* %	*Sample (c)* %
Volatile ether	2.33	3.84	4.03
Fixed ether	4.27	5.10	5.49
Total ether	6.60	8.94	9.52

Sample (a) was a poor quality allspice, long in storage; sample (b) was a fair quality product; sample (c) was a good quality allspice recently imported. The total ash of sample (c) was 4.45%.

Anise

Anise fruit contains a volatile oil, fixed oil, protein, cellulose, sugar, pentosans, pigment, calcium oxalate, mineral elements, etc.

United States Government Standards require that the dried fruit of anise shall contain not more than 9% of total ash, nor more than 1.5% of ash insoluble in hydrochloric acid."

The important constituent of anise is its volatile oil, present in an amount of from about 1.5 to 3.5%. It is a colorless or pale yellow liquid, slightly soluble in water, soluble in about 3 volumes of 90% alcohol, soluble in chloroform, and in ether.

The principal constituent of the volatile oil of anise is anethole, empirical formula, $C_{10}H_{12}O$, structural formula, $CH_3CH{:}CHC_6H_4OCH_3$, present in the amount of from 80 to 90%. Anethole crystallizes from oil of anise on cooling. In the pure state, above 23°C, anethole is a colorless, highly refractive liquid possessing the aromatic odor and sweet taste of anise.

Other organic compounds reported present in the volatile oil of anise include methyl chavicol, empirical formula, $C_{10}H_{12}O$; anisketone, $C_{10}H_{12}O_2$; and anisaldehyde, $C_8H_8O_2$.

S.R. El-Deeb, M.S. Karawya, S.K. Wahba, and F.M. Ebeid (Fac. Pharm., Univ. Cairo) report that by gas-liquid partition chromatography oil of anise has the peaks of anethole and anisaldehyde.*

The fixed oil content of anise is about 10%.

Bay Leaves

See *Laurel Leaves*.

Basil

See *Sweet Basil*.

---

* J. Pharm. Sci. (U. Arab. Rep.)3(1), 63-79 (1962); *CA*. 61, 8. 9351a

Capsicum

Capsicum fruits contain a fixed oil, resin, coloring matter, a pungent principle, protein, cellulose, pentosans, mineral elements, etc.

United States Government Standards cover paprika and cayenne pepper, the spice products of the capsicum fruits with which we are concerned, as follows: "Paprika. The dried, ripe fruit of *Capsicum annuum* L. It contains not more than 8.5% of total ash, not more than 1% of ash insoluble in hydrochloric acid. The iodine number of its extracted oils is not less than 125 nor more than 136.

"Hungarian Paprika. Paprika having the pungency and flavor characteristic of that grown in Hungary.

(a) Rosenpaprika, rosapaprika, rose paprika, is Hungarian paprika prepared by grinding specially selected pods of paprika, from which the placentae, stalks, and stems have been removed. It contains no more seeds than the normal pods, not more than 18% of nonvolatile ether extract, not more than 23% of crude fiber, not more than 6% of total ash, nor more than 0.4% of ash insoluble in hydrochloric acid.

(b) Koenigspaprika, king's paprika, is Hungarian paprika prepared by grinding whole pods of paprika without selection, and includes the seeds and stems naturally occurring with the pod. It contains not more than 18% of nonvolatile ether extract, not more than 23% of crude fiber, not more than 6.5% of total ash, nor more than 0.5% of ash insoluble in hydrochloric acid.

"Pimenton, Pimiento, Spanish Paprika. Paprika having the characteristics of that grown in Spain." It contains not more than 18% of nonvolatile ether extract, not more than 21% of crude fiber, not more than 8.5% of total ash, nor more than 1 per cent of ash insoluble in hydrochloric acid.

"Cayenne Pepper, Cayenne. The dried, ripe fruit of *Capsicum frutescens* L., *C. baccatum* L., or some other small-fruited species of capsicum. It contains not less than 15% of nonvolatile ether extract, not more than 1.5% of starch, not more than 28% of crude fiber, not more than 8% of total ash, nor more than 1.25% of ash insoluble in hydrochloric acid."

Canadian Government Standard, B.07.014: "CAYENNE PEPPER or CAYENNE, whole or ground, shall be the dried, ripe fruit of *Capsicum frutescens* L., *Capsicum baccatum* L., or other small-fruited species of Capsicum, and shall contain not more than

(a) 1.5% starch.

(b) 28% crude fiber.

(c) 8% total ash.

(d) 1.25% ash insoluble in hydrochloric acid, and shall contain not less than 15% non-volatile ether extract."

B.07.015: "PAPRIKA, whole or ground, shall be the dried, ripe fruit of *Capsicum annuum* L., and shall contain not more than

(a) 18% non-volatile ether extract.
(b) 23% crude fiber.
(c) 8.5% total ash.
(d) 1% ash insoluble in hydrochloric acid."

The National Formulary Standard states: "Capsicum is the dried ripe fruit of *Capsicum frutescens* Linn., known in commerce as African chillies, or *Capsicum annuum* Linne var. conoides Irish, known in commerce as Louisiana Long Pepper, or, of a hybrid between the Honka variety of Japanese Capsicum and the old Louisiana Sport Capsicum, known in commerce as Louisiana Sport Pepper (Family *Solanaceae*). Capsicum must be labeled to indicate which of the above varieties is contained in the package. Capsicum contains not more than 3% of its stems and calyxes, and not more than 1% of other foreign organic matter. It yields not less than 12% of non-volatile ether-soluble extractive and not more than 1.25% of acid-insoluble ash."

The principal coloring matter of capsicum fruits is the carotenoid pigment capsanthin, $C_{40}H_{58}O_3$. Other pigments present are carotene, capsorubin, zeaxanthin, cryptoxanthin, and lutein.

The pungent principle of capsicum is capsaicin, or capsaicine, $C_{18}H_{27}O_3N$. In the pure state, capsaicin consists of white crystals possessing a burning taste and very acrid vapors; insoluble in water, but freely soluble in ether, alcohol, and alkalies.

The oleoresin of capsicum, obtained by ether, alcohol, or acetone extraction, consists of a thick liquid, dark reddish-brown in color, with an intensely pungent taste. The chief active principle is capsaicin.

The quantity of capsaicin present in capsicum fruits varies, being greatest in the hot peppers, and slight or almost nil in the mild or sweet fruits.

Capsicum peppers contain appreciable quantities of Provitamin A, and Vitamins C, P, and J.*

Caraway

Caraway fruit contains a volatile oil, fixed oil, protein, cellulose, pentosans,

---

*Instituto Nacional de Investigaciones Agronomicas Ministerio De Agricultura, Murcia, Spain. UNE 34029

Vitamin J—"A postulated antipneumonia principle which was also designated C_2." *Vitamins and Coenzymes*, Arthur Wagner and Karl Folkers, Copenhagen, 1964 John Wiley and Sons Inc., N.Y.

calcium oxalate, mineral elements, etc. **United States Government Standards** require that the dried fruit of caraway shall contain "not more than 8% of total ash, nor more than 1.5% of ash insoluble in hydrochloric acid."

Canadian Government Standard, B.07.019: "CARAWAY SEED shall be the dried fruit of the caraway plant, and shall contain not more than

(a) 8% total ash.

(b) 1.5% ash insoluble in hydrochloric acid."

The principal constituent of caraway fruit is its volatile oil, of which it contains from about 3 to 7.5%. The chief constituents of the volatile oil of caraway are a ketone, d-carvone, empirical formula, $C_{10}H_{14}O$, and a terpene, d-limonene, $C_{10}H_{16}$. Many other organic compounds are reported present in the oil including dihydrocarvone, dihydrocarveol, carveol, acetaldehyde, methyl alcohol, furfural, diacetyl, etc. Carvone is present in the oil in the amount of from about 50 to 60%. In the pure state, carvone is a colorless liquid, very slightly soluble in water, soluble in alcohol, and in ether; it is primarily responsible for the caraway flavor. Limonene adds a lemon-like note to the oil; in the pure state it is a colorless liquid, insoluble in water, soluble in alcohol, and in ether.

S. R. El-Deeb, M. S. Karawya, S. K. Wahba, and F. M. Ebeid (Fac. Pharm., Univ. Cairo) report that by gas-liquid partition chromatography oil of caraway had peaks methyl salicylate, limonene, carvone, eugenol, α-pinene, and eucalyptol.*

The fixed oil content of caraway is about 15%. The fatty acids consist largely of oleic, linoleic, and petroselinic acids, with a small amount of palmitic acid. The discoverers of petroselinic acid showed it to be a higher-melting isomer of oleic acid.†

Cardamom

Cardamom seed contains a volatile oil, fixed oil, protein, cellulose, pentosans, sugar, starch, silica, calcium oxalate, mineral elements, etc.

United States Government Standards require that the dried seed of cardamom shall contain "not more than 8% of total ash, nor more than 3% of ash insoluble in hydrochloric acid."

Canadian Government Standard, B.07.020: "CARDAMOM SEED shall be the dried seed of cardamom, and shall contain not more than

(a) 10% total ash.

(b) 3% ash insoluble in hydrochloric acid."

*J. Pharm. Sci. (U. Arab Rep.) 3(1), 63-79 (1962); *CA.* 61,8. 9351a

†Eckey, E. W., *Vegetable Fats and Oils*, p. 707, Reinhold Publishing Corp., New York, 1954.

Cardamom contains from about 2 to 10% of volatile oil, colorless or pale yellow, with an agreeable, aromatic, somewhat camphoraceous odor, and a pungent, aromatic taste.

M. C. Nigam, Ishwar C. Nigam, K. L. Handa, and Leo Levi (Food and Drug Directorate, Ottawa, Canada) report that an examination of the essential oil of cardamom by the gas-chromatographic analysis of fractions obtained by distillation under reduced pressure and by column chromatography showed 1.9% a-pinene, 4.5% sabinene, 0.2% myrcene, 14.3% limonene, 30.7% cineole, 1.9% p-cymene, 0.8% methyl heptenone, 0.9% linalool, 1.2% linalyl acetate, 0.8% β-terpineol, 3.7% 2-terpineol, 28.1% a-terpinyl acetate, 0.1% borneol, 0.3% neryl acetate, 0.7% geraniol, 1.4% nerol, 0.3% neolidol, 0.5% heptacosane, and 7.7% unidentified compounds.*

The following approximate composition of Mysore, Malabar, and Ceylon cardamom oils is reported by Y. S. Lewis, E. Sankaran Nambudiri, and T. Philip (Central Food Technol. Res. Inst., Mysore, India)

Constituent	*Mysore*	*Malabar*	*Ceylon*
α-Pinene	1.4	0.7	0.7
Sabinene	3.2	4.6	3.8
Myrcene	0.2	0.2	0.1
Limonene	2.4	1.7	2.2
Cineole	41.0	26.5	36.0
p-Cymene	0.5	0.7	0.4
Methylheptanone	1.2	1.5	0.8
Linaloöl	0.4	3.7	3.5
Linalyl acetate	1.6	7.7	2.6
β-Terpineol	0.8	0.7	2.1
α-Terpineol	0.8	1.0	1.0
α-Terpinyl acetate	30.00	34.5	30.0
Neryl acetate (+borneol?)	1.1	1.2	0.8
Geraniol	0.7	0.7	0.7
Nerol	1.4	0.6	1.6
Nerolidol	0.3	0.7	0.3
Solid (heptacosane?)	0.5	0.4	0.5
Unidentified compounds	12.5	12.9	12.9

The yield of oil by steam distillation was Mysore 9.1%, Malabar 10.0%, Ceylon 9.5%.**

An interesting report on the yield of volatile oil in cardamom seed when exposed to air for various lengths of time is that of Ogzewalla and Marlene Willms (Southwestern State College, Weatherford, Okla. U.S.A.). They report that the yield of oil from freshly gathered whole cardamom fruit was 4.8%; exposed to air 1 month, 2.9%; exposed to air 6 weeks, about 2.4%;

* J. Pharm. Sci. 54(5), 799-801 (1965); *CA*. 63.2. 1651d.
** *Perfum. Essent. Oil Rec.* 57(10), 623-8 (1966) (Eng); *CA*. 66. 13993p.

exposed to air 14 months, 2.0%. Ground cardamom seed exposed to air 1 week, 2.4%; exposed to air 13 weeks, trace.*

Cassia

Cassia bark contains a volatile oil, fixed oil, resin, tannin, proteins, cellulose, pentosans, starch, calcium oxalate, mineral elements, etc.

United States Government Standards require that the powder made from cinnamon, in which they include the barks of Ceylon cinnamon, China cassia, and Saigon cassia, shall contain "not more than 5% of total ash, nor more than 2% of ash insoluble in hydrochloric acid."

Canadian Government Standard, B.07.007: "CINNAMON or CASSIA, whole or ground, shall be the dried bark of cultivated varieties of *Cinnamomum zeylanicum* Nees, or *C. cassia* L., from which the outer layers may have been removed, and shall contain not more than

(a) 5% ash.
(b) 2% ash insoluble in hydrochloric acid."

B.07.008: "CEYLON CINNAMON, whole or ground, shall be cinnamon obtained exclusively from *Cinnamomum zeylanicum* Nees."

The quantity of volatile oil and other ether-soluble organic compounds present in the spice depends upon the nature, quality, and condition of the bark. Employing the continuous extraction method, the author obtained the following results.

		Volatile Ether Extract %	*Fixed Ether Extract* %	*Total Ether Extract* %
Saigon Cassia,	sample (a)	4.72	3.85	8.57
	sample (b)	4.95	4.60	9.55
Batavia Cassia,	sample (a)	4.89	1.14	6.03
	sample (b)	4.50	1.15	5.65
Yunan Cassia,	sample (a)	2.87	2.47	5.34
	sample (b)	3.07	2.40	5.47
Kwangsi Cassia,	sample (a)	2.20	2.09	4.29
	sample (b)	2.21	1.89	4.10
China Cassia Chips		1.77	2.10	3.87

The total ash was determined by the author in five samples of cassia as follows:

		Total Ash %
Saigon Cassia,	sample (a)	4.29
	sample (b)	3.88
Batavia Cassia,	sample (b)	3.25
Yunan Cassia,	sample (a)	2.26
	sample (b)	2.20

* *Proc. Okla. Acad. Sci.* 43, 57-60 (1962) (Pub 1963); *CA.* 60.7. 7867b.

The volatile oil of cassia contains from 80 to 95% of cinnamic aldehyde (described under Ceylon cinnamon). Other organic compounds reported present in the oil include cinnamyl acetate, and very small amounts of cinnamic acid, benzaldehyde, methyl salicylaldehyde, methyl *ortho*-coumaraldehyde, salicyaldehyde, salicylic acid, and courmarin.

Cassia Buds

Cassia buds contain a volatile oil, fixed oil, proteins, cellulose, pentosans, starch, calcium oxalate, mineral elements, etc.

The chief constituent of the volatile oil is cinnamic aldehyde, empirical formula C_9H_8O.

Celery

Celery fruit contains a volatile oil, fixed oil, resin, protein, cellulose, calcium oxalate, mineral elements, etc.

United States Government Standards require that the dried fruit of celery shall contain "not more than 10% of total ash, nor more than 2% of ash insoluble in hydrochloric acid."

Canadian Government Standard, B.07.021: "CELERY SEED shall be the dried fruit of the celery plant, and shall contain not more than

(a) 10% total ash.

(b) 2% ash insoluble in hydrochloric acid."

B.07.029: "CELERY SALT shall be a combination of ground celery seed or ground dehydrated celery and salt and shall contain not more than 75% salt."

B.07.030: "CELERY PEPPER shall be a combination of ground celery seed or ground dehydrated celery and ground black pepper and shall contain not more than 70% ground black pepper."

The principal constituent of celery fruit is its volatile oil, of which it contains from about 2 to 3%. The chief constituents of the oil are a terpene, *d*-limonene, empirical formula $C_{10}H_{16}$, a substance with a lemon-like odor, present in the amount of 60%, and a sesquiterpene, selinene $C_{15}H_{24}$, present in the amount of from 10 to 20%. The characteristic odor of this spice is believed to be mainly due to two oxygenated compounds present in the oil—namely, sedanolide $C_{12}H_{18}O_2$, and sedanonic acid anhydride $C_{12}H_{16}O_2$.

The fixed oil content of celery fruit is about 16%. The fatty acids consist largely of petroselinic, oleic, and palmitic acids.

Chervil

Chervil contains a volatile oil, cellulose, pigment, mineral elements, etc.

There does not appear to be any commercial production of the essential oil of chervil. It has probably been distilled in the laboratory and investigated but the literature seems to be silent concerning the chemical constituents of the oil.

Cinnamon, Ceylon

Ceylon cinnamon bark contains a volatile oil, fixed oil, tannin, resin, proteins, cellulose, pentosans, mucilage, starch, calcium oxalate, mineral elements, etc.

On distillation the bark yields from 0.5 to 1% of volatile oil. The chief constituent of the volatile oil of cinnamon bark is cinnamic aldehyde, empirical formula C_9H_8O, structural formula $C_6H_5CH{:}CH.CHO$, an aromatic, pungent, sweet-tasting, yellowish, oily liquid. Other organic compounds reported present in the oil include eugenol (4 to 8% or more), caryophyllene, *l*-phellandrene, *p*-cymene, *l*-α-pinene, *l*-linaloöl, furfural, methyl-*n*-amyl ketone, nonyl aldehyde, benzaldehyde, hydrocinnamic aldehyde, and cumaldehyde, all of which contribute to the very pleasing aroma and taste of Ceylon cinnamon.

Concerning the distillation and export of the volatile oil of cinnamon, the following is reported in Leaflet No. 51, Department of Agriculture, Ceylon, by F.A. Stockdale and J.C. Drieberg: "Cinnamon oil is distilled from the chips and trimmings of the quills; and this is known as 'bark oil.' " Before distillation the chips are macerated in sea water or a strong solution of brine for 2 to 3 days. The yield of oil is 0.5 to 1%. Eighty lb. newly prepared cinnamon yields about 5½ oz. of heavy oil and 2½ oz. light oil.

"It is unfortunate that the greater part of the cinnamon oil exported is not genuine bark oil. Either leaves are added to bark when distilled, or cinnamon leaf oil is added to bark oil after distillation. . . . The active constituent of bark oil is cinnamic aldehyde; whereas that of leaf oil is eugenol, which is also the characteristic chemical constituent of oil of cloves. Leaf oil is sometimes designated 'clove oil.' "

For United States Government and Canadian standards see under *Cassia*.

Cloves

Cloves contain a volatile oil, fixed oil, resin, tannin, proteins, cellulose, pentosans, mineral elements, etc.

United States Government Standards require that cloves shall contain "not more than 5% of clove stems, not less than 15% of volatile ether extract, not less than 12% quercitannic acid (calculated from the total oxygen absorbed by the aqueous extract), not more than 7% of total ash, nor more than 0.5%

of ash insoluble in hydrochloric acid."

Canadian Government Standard, B.07.001: "CLOVES, whole or ground, shall be the dried flower buds of the clove plant, and shall contain not more than

(a) 5% clove stems.
(b) 8% total ash.
(c) 0.5% ash insoluble in hydrochloric acid.
(d) 10% crude fiber.

and shall contain not less than 15% volatile ether extract."

Published analyses* of cloves indicate considerable variation in the quantities of volatile and nonvolatile ether extracts present in this spice; minimum and maximum percentages reported, amongst others, as follows.

			Volatile Ether Extract %	*Nonvolatile Ether Extract* %
Penang Cloves	(1)	Min.	14.8	9.5
		Max.	17.2	12.0
	(2)	Min.	17.99	5.12
		Max.	20.04	6.61
Amboyna Cloves	(1)	Min.	18.00	8.2
		Max.	19.2	10.0
	(2)	Min.	20.42	6.39
		Max.	20.53	6.67
Zanzibar Cloves	(1)	Min.	12.1	8.00
		Max.	18.3	10.7
	(2)	Min.	17.82	6.24
		Max.	18.32	6.59

An analysis of Madagascar cloves by the author gave the following results.:

	Volatile Ether Extract %	*Nonvolatile Ether Extract* %	*Total Ether Extract* %	*Total Ash* %
Sample (a)	16.14	5.99	22.13	6.47
Sample (b)	17.26	6.30	23.56	6.41

The volatile oil of cloves contains from 70 to 90% of eugenol, empirical formula $C_{10}H_{12}O_2$, graphic formula

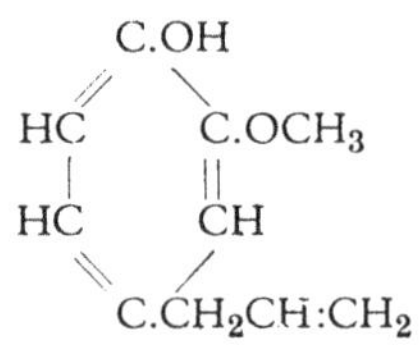

* Winton, *Structure and Composition of Foods*, Vol IV, p. 296, John Wiley & Sons, Inc., New York, 1939.

Eugenol is a colorless, or faintly yellow, optically inactive liquid with the aromatic odor and pungent taste of cloves; it becomes darker and thicker on exposure to air. Eugenol is very slightly soluble in water, miscible with alcohol, chloroform, ether, and fixed oils. The color and solubility of the volatile oil of clove is similar to that of eugenol.

Other organic compounds reported present in the oil of cloves include eugenol acetate, caryophyllene, methyl *n*-amyl carbinol, methyl *n*-heptyl carbinol, methyl *n*-heptyl ketone, methyl benzoate, benzyl alcohol, furfuryl alcohol, and vanillin.

S. R. El-Beeb, M.S. Karawya, S.K. Wahba, and F.M. Ebeid (Fac. Pharm., Univ. Cairo) report that by gas-liquid partition chromatography oil of cloves had peaks of eugenol, methyl salicylate, limonene, carvone, α-pinene, and eucalyptol.*

Coriander

Coriander fruit contains a volatile oil, fixed oil, tannin, protein, cellulose, pentosans, calcium oxalate, mineral elements, etc.

United States Government Standards require that the dried fruit of coriander shall contain "not more than 7% of total ash, nor more than 1.5 per cent of ash insoluble in hydrochloric acid."

Canadian Government Standard, B.07.022: "CORIANDER SEED shall be the dried fruit of the coriander plant and shall contain not more than

(a) 7% total ash.

(b) 1.5% ash insoluble in hydrochloric acid."

Coriander fruit contains from about .1 to 1% of volatile oil; it is a colorless or pale yellow liquid with the odor and flavor of the spice. The chief constituent of the oil is coriandrol (*d*-linalool), empirical formula $C_{10}H_{18}O$. In the pure state, coriandrol (*d*-linalool) is a colorless, fragrant liquid, very slightly soluble in water, soluble in alcohol, and in ether.

Other organic compounds reported present in the volatile oil of coriander include *d*-α-pinene, *dl*-α-pinene, β-pinene, β-terpinene, γ-terpinene, geraniol, borneol, and decylaldehyde.

G. M. Makarova and Yu. G. Borisyuk investigated the volatile oil of coriander and report linalool content 66.0%; and aldehyde content 0.2%. Fractional distillation separated 18% terpenes, such as α-pinene, β-pinene, dipentene, *p*-cymol, and α- and β-terpinenes.†

Yu. A. Akimov and V. G. Voronin report the following composition of coriander oil: α-pinene, camphene, β-pinene, myrcene, dipentene, *p*-

---

* J. Pharm. Sci. (U. Arab Rep.) 3(1), 63-79 (1962); *CA.* 61.8. 9351a.

† Farmatsevt, Za. (Kiev) 14, No. 3, 43-6 (1959); *CA.* 58.3.2320e.

cymene, linalool, camphor, linalyl acetate, borneol, geranyl acetate, and geraniol.*

The fixed oil content of coriander is about 20%. The fatty acids consist largely of petroselinic, oleic, palmitic, and linoleic acids.

Cumin

Cumin fruit contains a volatile oil, fixed oil, protein, cellulose, sugar, pentosans, calcium oxalate, mineral elements, etc.

United States Government Standards require that the dried fruit of cumin shall contain "not more than 9.5% of total ash, not more than 1.5% of ash insoluble in hydrochloric acid, nor more than 5% of harmless foreign matter."

Cumin fruit contains from about 2.5 to 4.5% of volatile oil. The chief constituent of the oil is cumaldehyde (*p*-isopropylbenzaldehyde; *p*-cuminic aldehyde), empirical formula $C_{10}H_{12}O$, structural formula $C_3H_7C_6H_4CHO$. Cumaldehyde, in the pure state, is a colorless to yellowish, oily liquid with a strong odor and flavor of cumin. It is insoluble in water, soluble in alcohol, and in ether.

Other organic compounds reported present in the volatile oil of cumin include dihydrocuminaldehyde, cuminyl alcohol, *dl*-pinene, *p*-cymene, and dipentene.

The fixed oil content of cumin is about 10%.

Curry Powder

Curry powder is a mixture or blend of a number of ground spices.

Canadian Government Standard, B.07.026: "CURRY POWDER shall be any combination of turmeric with spices and seasoning, and shall contain not more than 5% salt."

Dill

Dill fruit contains a volatile oil, fixed oil, protein, cellulose, calcium oxalate, mineral elements, etc.

United States Government Standards require that the dried fruit of dill shall contain "not more than 10% of total ash, nor more than 3 per cent of ash insoluble in hydrochloric acid."

Canadian Government Standard, B.07.023: "DILL SEED shall be the dried fruit of the dill plant, and shall contain not more than

(a) 10 per cent total ash, and

(b) 3 per cent ash insoluble in hydrochloric acid."

* *Gaz. Khromatogr.*, Moscow No. 3, 142-6 (1965) (Russ.); *CA*. 66. 21. 98424b.

Dill fruits contain from about 2 to 4% of volatile oil, colorless or pale yellow, and with the odor of the spice. The constituents of the oil are carvone, present in amount from about 40 to 50% or more, *d*-limonene, phellandrene, and probably other organic compounds. Carvone, empirical formula, $C_{10}H_{14}O$, a terpene ketone, is a colorless, oily liquid possessing an odor of caraway; it is very slightly soluble in water, soluble in alcohol, and in ether. The presence of carvone permits the use of dill as a substitute for caraway.

The fixed oil content of dill is about 17%.

Fennel

Fennel fruit contains a volatile oil, fixed oil, protein, cellulose, sugar, pentosans, calcium oxalate, mineral elements, etc.

United States Government Standards require that the dried fruit of fennel shall contain "not more than 9% of total ash nor more than 2% of ash insoluble in hydrochloric acid."

Fennel fruit contains from about 3 to 4% of volatile oil, colorless or pale yellow, with the characteristic odor and taste of the spice. The chief constituents of the oil are anethole, about 50 to 60%, and fenchone, about 20%. Anethole is described under anise; its anise-like odor is modified in fennel by the presence of fenchone and the other organic compounds present. Fenchone (*d*-fenchone), empirical formula $C_{10}H_{16}O$, a ketone, is an oily liquid with a bitter, camphoraceous flavor; insoluble in water, soluble in alcohol, and in ether.

Other organic compounds reported present in the volatile oil of fennel include *d*-α-pinene, *d*-α-phellandrene, dipentene, methyl chavicol, feniculin, anisaldehyde, and anisic acid.

I.U.W. Osisiogu (Univ. Nigeria, Nsukka) in a study of the fennel oil produced at Nsukka reported that the fruit of *Foeniculum vulgare*, of Indian origin and cultivated in Nigeria yielded 2.0-2.4% of essential oil, which contained $\sim$86% anethole but no fenchone.*

The fixed oil content of fennel is about 10%. The fatty acids consist largely of petroselinic, oleic, linoleic, and palmitic acids.

I. M. Fefer investigated the solid part of the greasy oil from fennel seeds and reports 1.18% unsaponifiables, 1.52% saturated glycerides, probably of palmitic and stearic acids, and 97.3% glycerides of petroselinic acid.**

Fenugreek

Fenugreek seed contains a volatile oil, fixed oil, resin, protein, cellulose, starch, sugar, mucilage, mineral elements, etc.

* *Planta Med.* 15(1), 30-1 (1967) (Eng); *CA.* 66. 24. 108169p

** *Farmatsevt. Zh.* (Kiev) 17, No. 1, 45-8 (1962); *CA.* 57.2.2354c.

The volatile oil content of fenugreek seed is very small, less than 0.02%. It is brown in color, and highly odorous. Very little is known about it.

The constituents of fenugreek seed include the alkaloids trigonelline, $C_7H_7O_2N$, and choline, $C_5H_{15}O_2N$, and a yellow coloring matter.

The fixed oil content is about 7%. The fatty acids consist largely of linoleic, oleic, and linolenic acids, with small amounts of palmitic, stearic, arachidic, and behenic acids.

Fenugreek seeds yield all of their odor and taste to alcohol.

Garlic

Garlic contains a volatile oil, fixed oil, protein, cellulose, mineral elements, etc.

The volatile oil content of garlic is very small, about 0.1%. It is a brownish-yellow liquid with a disagreeable, pungent odor. The constituents of the oil include diallyl disulfide, $C_6H_{10}S_2$ (about 60%), diallyl trisulfide, $C_6H_{10}S_3$ (about 20%), allyl propyl disulfide, $C_6H_{12}S_2$ (about 6%), a small quantity of diethyl disulfide, and probably diallyl polysulfide. Diallyl disulfide, a liquid, is said to possess the true garlic odor.

O. E. Schultz and H. L. Mohrmann, investigating the effectiveness of a number of mobile solvents for separating garlic oil constituents by thin-layer chromatography, report that between 9 and 14 oil constituents were separated. Diallyl disulfide was the main component of the oil.*

Canadian Government Standard, B.07.028: "GARLIC SALT shall be a combination of powdered dehydrated garlic and salt, shall contain not more than 75% salt, and may contain not more than 2% of an anti-caking agent."

Ginger

Ginger contains a small quantity of volatile oil, a fixed oil with pungent resinous matter, proteins, cellulose, pentosans, starch, mineral elements, etc.

United States Government Standards require that dried ginger shall contain "not less than 42% of starch, not more than 8% of crude fiber, not more than 1% of lime (CaO), not less than 12% of cold water extract, not more than 7% of total ash, not more than 2 per cent of ash insoluble in hydrochloric acid, nor less than 2% of ash soluble in cold water.

"Jamaica ginger. Ginger grown in Jamaica. It contains not less than 15% of cold water extract, and conforms in other respects to the standards for ginger.

---

* *Pharmazie* 20 (6), 379-81 (1965); *CA.* 63.7.8116f.

"Limed ginger, bleached ginger. Whole ginger coated with carbonate of calcium. It contains not more than 4% of carbonate of calcium nor more than 10% of total ash, and conforms in other respects to the standards for ginger."

Canadian Government Standard, B.07.002: "GINGER, whole or ground

(a) shall be the washed and dried or decorticated and dried rhizome of the ginger plant;
(b) shall contain not more than 10% moisture;
(c) shall contain on the dry basis, not less than
  (i) 45% ginger starch.
  (ii) 13.3% cold water extractive as determined by the official method.
  (iii) 2% ash soluble in water.
(d) may contain not more than
  (i) 9% crude fiber.
  (ii) 1% calcium, calculated as calcium oxide.
  (iii) 7.5% total ash.
  (iv) 2% ash insoluble in hydrochloric acid."

B.07.003: "JAMAICA GINGER, whole or ground, shall be ginger grown in Jamaica, and shall conform to the standard provided in section **B.07.002** except that it shall contain, on the dry basis, not less than 16.6% cold water extractive as determined by the official method."

B.07.004: "LIMED GINGER or BLEACHED GINGER, whole or ground, shall be ginger coated with calcium carbonate and shall conform to the standard provided in **B.07.002** except that it shall contain not more than

(a) 2% calcium, calculated as calcium oxide.
(b) 11% total ash."

The determination of cold water extract is for the purpose of detecting exhausted or spent ginger, that is, ginger which has been previously employed in the making of products containing the flavoring principles of natural ginger. It is not likely that the spice will consist entirely of exhausted ginger, but the standard is there to detect adulteration by the admixing of exhausted ginger with the pure spice.

The author obtained a result of 16.29% in a determination of cold water extract in a random sample of Jamaica ginger of good quality.

Ginger contains from 1 to 3% of volatile oil. Oil of ginger is a yellowish, viscid liquid soluble in ether, sparingly soluble in alcohol, and almost insoluble in water. The chief constituent of oil of ginger is a sesquiterpene, zingiberene, $C_{15}H_{24}$. Other organic compounds reported present in the oil include α- and β-phellandrene, *d*-camphene, methyl heptenone, *n*-decylaldehyde, *n*-nonylaldehyde, borneol, cineole, linalool, citral, and a sesquiterpene alcohol zingiberol, $C_{15}H_{26}O$, an odorous compound from which the characteristic gingerlike odor of the oil is derived.

M. C. Nigam, Ishwar C. Nigam, and Leo Levi (Food and Drug Directorate. Ottawa. Canada) and K. L. Handa report that oil of ginger contains the following, unreported, constituents: cumene, β-pinene, myrcene, *p*-cymene, 2-selinene, β-elemene, 2-heptanol, 2-nonanol, and bornyl acetate.*

Oleoresin. The pungent principle of ginger is zingerone, present in the oleoresin obtained from the spice by extraction with acetone, alcohol, or ether. Zingerone, 3-methoxy-4-hydroxy-phenylethylmethylketone, empirical formula $C_{11}H_{14}O_3$, graphic formula

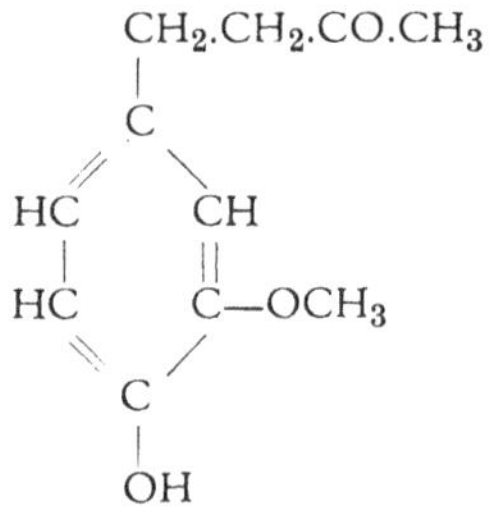

Consists, in the pure state, of white crystals, melting at 40 to 41° C; sparingly soluble in water, soluble in ether.

The oleoresin also contains shogaol, unidentified resins, and volatile oil. Shogaol, empirical formula $C_{17}H_{24}O_3$, is an unsaturated homologue of zingerone. The substance was obtained from ginger by Nomura of Japan who named it shogaol after "shoga", the Japanese name of ginger.

Laurel Leaves

Laurel leaves (Bay leaves) contain a volatile oil, fixed oil, protein, cellulose, pentosans, pigment, mineral elements, etc.

The principal constituent of the laurel leaf is its volatile oil, present in the amount of about 2%. The chief constituent of the oil is cineole, empirical formula $C_{10}H_{18}O$, a colorless liquid with a strong, aromatic, camphoraceous odor, and a cooling taste. Cineole is present in the oil in the amount of from about 45 to 50%. Other organic compounds reported present in the oil include α-pinene, α-phellandrene, *l*-linaloöl, *l*-α-terpineol, geraniol, eugenol, eugenol acetate, methyl eugenol, a number of esters, and acetic, isobutyric, and isovaleric acids.

---

* Can. J. Chem. 42 (11), 2610-15 (1964); *CA.* 61.12.14463a.

Mace

Mace contains a volatile oil, fixed oil, resin, cellulose, pentosans, amylo-dextrin-starch, mineral elements, etc.

United States Government Standards require that mace shall contain "not less than 20% nor more than 30% non-volatile ether extract, not more than 10% of crude fiber, not more than 3% of total ash, nor more than 0.5% of ash insoluble in hydrochloric acid."

Canadian Government Standard, B.07.009: "MACE, whole or ground, shall be the dried arillus of *Myristica fragrans* Houttyn and shall contain not more than

(a) 7% crude fiber;
(b) 3% total ash.
(c) 0.5% ash insoluble in hydrochloric acid.

and the non-volatile ethyl ether extract, obtained after extraction of mace with petroleum ether, shall not exceed 5% and the sum of the nonvolatile extracts with petroleum ether and ethyl ether shall not exceed 33%."

The volatile oil is similar to that of nutmeg (page 200).

Marjoram

See *Sweet Marjoram.*

Mustard

White and black mustard seed contains a fixed oil, protein, cellulose, pentosans, mucilage, mineral elements, etc.

United States Government Standards for white and black mustard seed require that Sinapsis alba (white mustard) shall contain "no appreciable amount of volatile oil. It contains not more than 5% of total ash nor more than 1.5% of ash insoluble in hydrochloric acid.

"Brassica nigra (black mustard) . . . yield 0.6% of volatile mustard oil (calculated as allyl isothiocyanate). . . . not more than 5% of total ash, nor more than 1.5% of ash insoluble in hydrochloric acid."

Canadian Government Standard, B.07.024: "MUSTARD SEED shall be the seed of *Brassica Bois; B. hirta* Moench; *B. nigra* (L.) Koch; *B. juncea* (L.) Czern; or seed of species closely related to *B. nigra* and *B. juncea*, and shall contain

(a) not more than 1.5% ash insoluble in hydrochloric acid;
(b) not more than 8% total ash, on the oil-free basis."

B.07.005: "MUSTARD, MUSTARD FLOUR or GROUND MUSTARD shall be the powder made from mustard seed with the hulls largely removed, from which a portion of the fixed oil may be removed, and shall contain

(a) not more than 1.5% starch;
(b) not more than 8% total ash, on the oil-free basis; and shall yield
(c) not less than 0.40% volatile mustard oil as determined by the official method."

The expressed oil of mustard is a straw-colored or brownish-yellow, or greenish-brown liquid; it is present in the amount of about 24 to 26% in white mustard and of about 24 to 40% in black mustard. In white mustard oil the fatty acids consist largely of erucic and oleic acids, with small amounts of linoleic, linolenic, eicosenoic, behenic, palmitic, lignoceric, arachidic, stearic, and myristic acids. Black mustard oil has similar constituents but without the glyceride of stearic acid.

The most important constituent of white mustard seed is the glucoside, sinalbin. In the presence of water, sinalbin is broken down or decomposed by the action of an enzyme, myrosin, yielding sinalbin mustard oil, d-glucose, and sinapine bisulfate. The reaction is said to proceed as follows:

$$C_{30}H_{42}O_{15}N_2S_2 + H_2O \longrightarrow C_7H_7ONCS + C_6H_{12}O_6 + C_{16}H_{24}NO_5HSO_4$$

sinalbin	sinalbin mustard oil (*p*-hydroxybenzyl isothiocyanate)	*d*-glucose	sinapine acid sulfate

Sinalbin mustard oil is only very slightly volatile with steam (sometimes reported as nonvolatile). It is almost odorless but has a pungent taste and causes blisters on the skin.

In black mustard seed the most important constituent is the glucoside sinigrin, or potassium myronate. When this substance is acted on by the enzyme myrosin in the presence of water it yields allyl isothiocyanate, *d*-glucose, and potassium hydrogen sulfate:

$$C_{10}H_{16}KNS_2O_9 + H_2O \longrightarrow CH_2{:}CH.CH_2.NCS + C_6H_{12}O_6 + KHSO_4$$

potassium myronate	allyl isothiocyanate	*d*-glucose	potassium hydrogen sulfate

Allyl isothiocyanate is a volatile liquid, known as volatile oil of mustard, with a pungent, irritating odor, and an acrid taste. It causes blisters on the skin.

Nutmeg

Nutmeg contains a volatile oil, fixed oil, proteins, cellulose, pentosans, starch, mineral elements, etc.

The volatile oil is obtained by distillation, the fixed or expressed oil by pressure accompanied by heat.

United States Government Standards require that nutmeg shall contain "not less than 25% of nonvolatile ether extract, not more than 10% of crude fiber, not more than 5% of total ash, nor more than 0.5% of ash insoluble in hydrochloric acid."

Canadian Government Standard, B.07.010: "**NUTMEG,** whole or ground, shall be the dried seed of *Myristica fragrans* Houttyn, may have a thin coating of lime, shall contain not less than 25% non-volatile ether extract, and shall contain not more than

(a) 5% total ash.

(b) 0.5% ash insoluble in hydrochloric acid."

In a published table* of the composition of spices the following values for ether extract are given for nutmeg of unstated origin.

	Volatile Ether Extract %	*Nonvolatile Ether Extract* %
Minimum	2.56	28.73
Maximum	6.94	36.94

Ether extract determinations, using the continuous extraction method, made by the author on ground Grenada nutmeg gave the following results.

	Volatile Ether Extract %	*Nonvoratile Ether Extract* %	*Total Ether Extract* %
Sample (a)	3.13	38.54	41.67
Sample (b)	3.19	38.85	42.04

The figures shown for volatile ether extract in the latter determinations are undoubtedly low. Ground nutmeg suffers a rapid loss of volatile substances at ordinary temperatures; furthermore, the long period of ether volatilization at room temperature, and dessicator drying, following extraction is perhaps not the best process for this spice.

The volatile oil of nutmeg, also called oil of myristica, is a colorless or pale yellow liquid with the odor and taste of nutmeg. The oil is reported to contain a number of organic compounds including *d*- and *l*-α-pinene, β-pinene, *d*-camphene, dipentene, *p*-cymene, *d*-linaloöl, safrole, *dl*-α-terpineol, borneol, geraniol, and myristicin. The last mentioned substance is toxic and said to act as a narcotic.*

* Winton, *Analysis of Foods*, 887, John Wiley & Sons, Inc., New York, 1945.

Ginette Jauregniberry and Robert Wolf (C.N.R.S., Gif-sur-Yvette, France) report the presence of (+)-sabinene in oil of nutmeg.*

Elwina A. Bejnarowicz and Ernst R. Kirch (Univ. of Illinois, Chicago) report that three East Indian and one West Indian varieties of nutmeg oil were analyzed by gas chromatography and that all four conatined α-pinene, β- pinene, limonene, *p*-cymene, 4-terpinenol, α-terpineol, and an unknown substance. Small amounts of terpinolene were found in two samples of East Indian nutmeg oil. Safrole and eugenol were present in all samples of East Indian nutmeg oil but not in West Indian nutmeg oil. Camphor was found only in West Indian nutmeg oil.**

Nutmeg yields from about 24 to 30% of fixed oil or expressed oil of nutmeg, also known as nutmeg butter, oil of mace. It is an orange-red to reddish-brown substance with the consistency of butter at ordinary temperatures. The chief constituent of the oil is trymyristin, $C_{45}H_{86}O_6$, the triglyceride of myristic acid. The remainder consists of the glycerides of oleic and linoleic acids, unsaponifiable matter, volatile oil, and resins.

Alexander T. Shulgin (Dow Chem. Co., Walnut Creek, Calif. U.S.A.) reports that the myristicin fraction of nutmeg oil after isomerization with alc. KOH was separated by gas chromatography to give methyl isoeugenol, *trans*-isomyristicin, and *cis*-isomyristicin.†

Onion

Onion contains a volatile oil, fixed oil, protein, cellulose, sugars, mineral elements, etc.

The volatile oil content of onion is very small, and little is known about its chemistry. The chief constituent of the oil is a disulfide with the empirical formula $C_6H_{12}S_2$. Other compounds containing sulphur may be present.

Canadian Government Standard, B.07.027: "ONION SALT shall be a combination of powdered dehydrated onion and salt, shall contain not more than 75% salt, and may contain not more than 2% of an anti-caking agent."

Origanum

Origanum contains a volatile oil, fixed oil, proteins, cellulose, pigment, mineral elements, etc.

This herb is usually classified as *Origanum vulgare* L. (common or wild mar-

* *Bull. Soc. Chim.* France 1962, 1958-7; *CA.* 58.3.2318d.

** *J. Pharm. Sci.* 52(10). 988-93 (1963); *CA.* 60.1. 364a.

† Nature 297, 379 (1963); *CA.* 58.9.8846h.

joram) but morphologically it differs from the published descriptions of wild marjoram; it may be a variety of *Origanum vulgare*, or an entirely different species. The plant owes its usefulness as a culinary herb to its volatile oil.

In the literature concerning the volatile oil of origanum it would appear that oils made under this name are derived from a number of related species of Labiatae plants. In some origanum oils, thymol would appear to be the main constituent; in others, carvacrol. It is probable that some quantity of each of these compounds is present in the volatile oil of the herb under discussion, together with terpenes, alcohols, esters, etc.

Parsley

Parsley leaves contain a volatile oil, cellulose, pigment, mineral elements, etc.

The chemistry of the fruits of parsley seems to be fairly well known, but little is known about that of the leaves. The fruits of parsley are not employed as a spice. However, all parts of the parsley plant contain a volatile oil, and it is likely that the oil of the leaves differs little, if at all, from the oil of the fruit.

The herb owes its odor and taste to its volatile oil, which probably has for its main constituent a large proportion of the phenol ether, apiole. Apiole, structural formula $CH_2{:}CHCH_2C_6H(OCH)_2(CH_2O_2)$, known also as parsley camphor, crystallizes in the form of long needles; it is colorless and has a faint parsley odor. Myristicin, α-pinene, and small amounts of aldehydes, ketones, and phenols, are present in the fruit oil, and these compounds probably exist in the leaf oil.

Pepper

Pepper contains a volatile oil, fixed oil, resin, alkaloids, proteins, cellulose, pentosans, starch, mineral elements, etc.

United States Government Standards require that black pepper shall contain "not less than 6.75% of nonvolatile ether extract, not less than 30% of starch, not more than 7% of total ash, nor more than 1.5% of ash insoluble in hydrochloric acid." The standards also require that white pepper shall contain "not less that 7% of nonvolatile ether extract, not less than 52% of starch, not more than 5% of crude fiber, not more than 3.5% of total ash, nor more than 0.3% of ash insoluble in hydrochloric acid."

Canadian Government Standard, B.07.011: "**BLACK PEPPER or PEPPERCORN,** whole or ground,

(a) shall be the dried, immature berry of *Piper nigrum* L.;

(b) shall contain not less than

(i) 6% non-volatile ether extract,
(ii) 30% pepper starch;
(c) may contain not more than
(i) 6% total ash.
(ii) 0.9% ash insoluble in hydrochloric acid;
(d) in the case of ground black pepper shall contain all the parts of the berry in their normal proportions."

B.07.013: "**WHITE PEPPER,** whole or ground,
(a) shall be the dried, mature berry of *Piper nigrum* L. from which the outer coating or the outer and inner coatings are removed
(b) shall contain not less than
(i) 7% nonvolatile ether extract,
(ii) 52% pepper starch;
(c) may contain not more than
(i) 5% crude fiber,
(ii) 2.2% total ash,
(iii) 0.3% ash insoluble in hydrochloric acid."

The author obtained the following results for ether extract determination:

	Volatile Ether Extract %	*Nonvolatile Ether Extract* %	*Total Ether Extract* %
Black pepper (origin unknown)	1.55	8.77	10.32
Black pepper (India)	2.05	8.79	10.84

And for ash, the following results:

	Total Ash %	*Ash Insoluble in Hydrochloric Acid* %
Black pepper (origin unknown)	6.39	1.49
Black pepper (India)	(a) 5.12	0.35
	(b) 5.19	not determined
	(c) 5.28	not determined

Pepper contains from 1 to 3% of volatile oil. It is a colorless or yellowish liquid; not pungent, but with the aromatic flavor of pepper; insoluble in water, soluble in about 10 to 15 parts of 90% alcohol, and in 3 to 10 volumes of 95% alcohol.

Organic compounds present in the volatile oil of pepper include α-pinene, β-pinene, *l*-α-phellandrene, β-caryophyllene, dihydrocarveol, and piperonal.*

C. J. Muller and W. G. Jennings (Univ. of California, Davis) identified the following as major compounds of the sesquiterpene hydrocarbon fraction of black pepper oil: δ-elemene, α-copaene, β-elemene, α-*cis*-bergamotene, α-santalene, α-*trans*-bergamotene, β-caryophyllene, α-humulene, β-selinene, α-selinene, β-bisabolene, δ-cadinene, and calamenene.**

The pungent principles of pepper are contained in its oleoresin, and consist of piperine, chavicine, piperidine. Piperine $C_{17}H_{19}NO_3$, an alkaloid, forms colorless, monoclinic needles from alcohol, at first tasteless but with a pungent aftertaste. Chavicine, $C_{17}H_{19}NO_3$, an isomer of piperine, is a yellowish, oily substance with a sharp, biting taste. Piperidine $(CH_2)_5NH$, an alkaloid, is a colorless liquid with the odor of pepper and a biting taste. To these, there has been added piperettine, a homologue of piperine, $C_{19}H_{21}O_3N$.†

Peppermint

Peppermint contains a volatile oil, fixed oil, resin, tannin, cellulose, pentosans, pigment, mineral elements, etc.

Oil of peppermint is a colorless to pale yellow liquid with a strong peppermint odor and pungent taste. It is very slightly soluble in water, soluble in 4 volumes of 90% alcohol.

The volatile oil of peppermint contains a large number of organic compounds. The chief constituents of the oil are *l*-menthol $C_{10}H_{20}O$, a secondary alcohol with a strong peppermint-like odor and a cooling taste; and *l*-menthone $C_{10}H_{18}O$, a ketone with an odor reminescent of peppermint, and a bitter taste less cooling than that of menthol. Other organic compounds present in the oil include terpinene, α-pinene, *l*-limonene, phellandrene, cadinene, *d*-menthone, menthyl acetate, menthyl isovalerate, amyl alcohol, acetaldehyde, isovaleric aldehyde, dimethyl sulfide, etc.

Poppy Seed

Poppy seed contains fixed oil, proteins, cellulose, pentosans, calcium oxalate, mineral elements, etc.

The most important constituent of poppy seed is the fixed oil, present in the amount of from 44 to 50%. It is straw- or pale-yellow in color, almost odor-

---

* Hewitt, Ritter, Konigsbacher and Hasselstrom, paper presented at the 12th International Congress for Pure and Applied Chemistry, N. Y. C., Sept. 1951. (*The Essential Oils*, Ernest Guenther, 5, 143.)

** *J. Agr. Food Chem.* 15 (5), 762-6 (1967) (Eng); *CA* 67. 20. 93905v

† F. S. Spring and J. Stark, *CA* 44, 22, 10685f, 1950.

less, and agreeable in taste. The fatty acids consist largely of linoleic and oleic acids with smaller amounts of palmitic and stearic acids.

Rosemary

Rosemary contains a volatile oil, fixed oil, resin, tannin, proteins, cellulose, pentosans, pigment, mineral elements, etc.

Rosemary yields from about 0.3 to 2% of volatile oil, a colorless or pale yellow liquid with the odor of the herb, and a warm, camphoraceous taste. Organic compounds reported present in the oil include borneol (10 to 15 per cent) pinene, camphene, cineole, camphor, esters, etc.

Borneol (bornyl alcohol; 2-hydroxy camphane; 2-camphanol) empirical formula $C_{10}H_{18}O$, an alcohol, forms colorless crystals possessing a pungent, camphor-like odor and a burning taste.

E. R. El-Deeb, M. S. Karawya, S. K. Wahba, and F. M. Ebeid (Fac. Pharm., Univ. Cairo) report that by gas-liquid partition chromatography oil of rosemary had the peaks of camphene, α-pinene, bornyl acetate, and eucalyptol.*

Safflower (Mexican or American Saffron)

Safflower contains two coloring matters—namely, safflor-yellow and carthamin.

Safflor-yellow, empirical formula $C_{24}H_{30}O_{15}$, is the yellow coloring matter of safflower, and present in the amount of from 24 to 30%; it is soluble in water and in alcohol. Carthamin, also known as safflor-red, empirical formula $C_{21}H_{22}O_{11}$, a glucoside, is present in the amount of from 0.3 to 0.6. In the pure state, carthamin is a dark red, granular powder with a green luster; it is slightly soluble in water, soluble in dilute carbonate solutions, and insoluble in ether.

Saffron

Saffron contains a volatile oil, fixed oil, protein, cellulose, pentosans, an enzyme, mineral elements, etc.

United States Government Standards require that the dried stigmas of saffron shall contain "not more than 10% of yellow styles and other foreign matter, not more than 14 per cent of volatile matter when dried at 100°C, not more than 7.5% total ash, nor more than 1% of ash insoluble in hydrochloric acid."

Saffron contains from about 0.5 to 1% of volatile oil, a yellow, limpid, mobile liquid with an intense odor of saffron. The oil contains a glucoside, pic-

* *J. Pharm. Sci.* (U. Arab Rep.) 3(1), 63-79 (1962); *CA.* 61.8.9351a.

rocrocin (saffron bitter) $C_{16}H_{26}O_7$, which yields on hydrolysis safranal and *d*-glucose.

The coloring matter consists of crocin $C_{44}H_{64}O_{28}.H_2O$, a glucoside, which yields on hydrolysis crocetin and d-glucose. In the pure state, crocin consists of brown-red crystals, slightly soluble in cold water, freely in hot water; almost insoluble in absolute alcohol, ether, and in most common organic solvents. Crocetin is a dark red powder nearly insoluble in water, soluble in ether, alcohol, and in alkaline solutions.

Sage

Sage contains a volatile oil, fixed oil, resin, tannin, proteins, cellulose, pentosans, pigment, mineral elements, etc.

Canadian Government Standard, B.07.017: "**SAGE,** whole or ground, shall be the dried leaves of the sage plant, and shall contain not more than 12% stems (excluding petioles) and other foreign material."

Sage yields up to about 2.5% of volatile oil. The chief constituents of the oil are thujone and borneol. Thujone, empirical formula $C_{10}H_{16}O$, is a colorless, odorous liquid, responsible for the strong, characteristic smell of sage.

Other organic compounds reported present in the oil include salvene, *d*-α-pinene, cineole, and *d*-camphor.

Carl Heinz Brieskorn and Siegfried Dalferth (Univ. Wuerzburg, Germany) report that in the distillation of oil of sage of Dalmatian origin through a 35-cm Vigreux column (30 mm Hg, reflux ration 1:30) the fraction b. 130-53° was rich in 2-methyl-3-methylene-5-heptene. Its possible biogenesis from thujone is discussed by the authors.*

Savory

Savory contains a volatile oil, fixed oil, proteins, cellulose, pentosans, sugar, pigment, mineral elements, etc.

Savory yields about 1% of volatile oil. The chief constituent of the oil is carvacrol, empirical formula, $C_{10}H_{14}O$, a colorless to pale yellow liquid with a pungent, thymol odor (it is isomeric with thymol).

Other organic compounds present in the volatile oil of savory include *p*-cymene, and probably pinene and dipentene.

Sesame

Sesame seed contains fixed oil, proteins, cellulose, pentosans, calcium oxalate, mineral elements, etc.

* *Ann.* 676. 171-4 (1964); *CA.* 61.12.14463d.

The chief constituent of sesame seed is the fixed oil, present in the seed to the extent of 57% or more, but usually from about 44 to 54%. The fatty acids consist largely of oleic and linoleic acids with small amounts of palmitic, stearic, arachidic, and myristic acids.

The unsaponifiable matter includes sesamin, empirical formula $C_{20}H_{18}O_6$, and sesamolin $C_{20}H_{18}O_7$, the former a complex cyclic ether, the latter with the characteristics of a glycoside.

Sesame oil is pale yellow in color, almost odorless, and possesses a bland taste. It is insoluble in water, slightly soluble in alcohol, and soluble in chloroform and ether.

Spearmint

Spearmint contains a volatile oil, fixed oil, resin, tannin, cellulose, pentosans, pigment, mineral elements, etc.

The volatile oil of spearmint is a colorless, yellow or greenish-yellow liquid with the characteristic spearmint odor and taste. It is very slightly soluble in water, soluble in an equal volume of 80% alcohol. The oil contains *l*-carvone (50% or more); *l*-limonene, *l*-phellandrene, and probably other terpenes; esters, etc.

The ketone, *l*-carvone, empirical formula $C_{10}H_{14}O$, in the pure state, is a colorless liquid with an odor of caraway, in which fruit it also occurs.

Thomas F. Burks and Gunnar Gjerstad (Univ. of Texas, Austin, U.S.A.) investigating the components of spearmint oil report that the presence of cineole, a-pinene, and linalool was conclusively demonstrated.*

Star Anise

Star anise contains a volatile oil, fixed oil, resin, protein, cellulose, pentosans, mineral elements, etc.

United States Government Standards require that star anise shall contain "not more than 5% of total ash."

Star anise yields about 3 % of volatile oil, colorless or pale yellow, with an aromatic odor, and a sweet, agreeable taste.

The chief constituent of the oil is anethole (described under anise); it constitutes from about 85 to 90 % of the oil. A large number of other organic compounds are reported present in the volatile oil of star anise including α- and β-phellandrene, *d*-α-pinene, *p*-cymene, *l*-limonene, α-terpineol, methyl chavicol, hydroquinone, monoethyl ether, safrole, anisketone, feniculin, and anisoxide.

* *J. Pharm. Sci.* 53(8), 964-5 (1964); *CA.* 61.10. 11844f.

Friedrich Porsch, Heinz Farnow, and Horst Winkler, investigating a specimen of the volatile oil from Hanoi, North Vietnam, report the presence of 88% *trans*-anethole and 5% (−)-limonene. Other constituents, present in amounts of 0.1 to 1%, were 8-hydroxy-*p*-cymol, terpinolene, myrcene, *l*-terpineol, (+)-α-terpineol, (−)-β-pinene, (+)-α-pinene. Δ^3-carene, a-phellandrene, cineole, and linalool. Traces of camphene, *p*-cymol and *l*-methyl-4-isopropenylbenzene were also determined. They observe that "Most of these compounds are reported for the first time; conversely, a number of compounds reported by other authors were not identified."*

Star anise contains about 20% of fixed oil; it contains the glycerides of stearic, linoleic, palmitic, and oleic acids.

Sweet Basil

Sweet basil contains a volatile oil, fixed oil, proteins, cellulose, pigment, mineral elements, etc.

The volatile oil of sweet basil is yellowish to greenish in color. The odor is aromatic. The chief constituents of the oil are methyl chavicol and *d*-linaloöl. Of the other organic compounds present in the volatile oil of sweet basil cineole and eugenol have been identified.

Methyl chavicol, empirical formula $C_{10}H_{12}O$, a phenol ether, an isomer of anethole, is a colorless liquid with an anise-like odor. Linaloöl, empirical formula $C_{10}H_{18}O$, an alcohol, is a liquid with a very pleasant odor reminiscent of lily of the valley. Cineole has a camphor-like odor, and eugenol a clove-like odor.

D. Ivanov, S. Chorbadzhiev, and Ch. Iordanov (Univ. Sofia, Bulg.), investigating the composition of Bulgarian basil oil, report that the Bulgarian basil resembles closely the American-type oil of sweet basil, containing linaloöl, methyl chavicol, 1,8-cineole, eugenol, geraniol, *p*-cymene, myrcene, and a trace of cinnamic acid.**

D. Ivanov, S. Chorbadzhiev, and Ch. Iordanov report the following constituents in the composition of Bulgarian basil oil: l-linaloöl (main constituent), 1,8-cineole, eugenol, geraniol, *p*-cymene, myrcene, α-terpineole, and sesquiterpene hydrocarbons.†

Sweet Marjoram

Sweet marjoram contains a volatile oil, fixed oil, proteins, cellulose, pentosans, pigment, mineral elements, etc.

* *Dragoco Rept.* 12(2), 27-9 (1965); *CA.* 63.1. 428d.

** *Perfumery Essential Oil Record* 55(11), 717-19 (1964); *CA.* 62.12.14418f.

† *God. Sofi. Univ., Khim. Fak.* 59 229-36, (1964-1965) (Pub. 1966) (Bulg); *CA.* 67. 23. 111332n.

United States Government Standards require that the dried leaves, with or without a small proportion of the flowering tops, of sweet marjoram shall contain "not more than 16% of total ash, not more than 4.5% of ash insoluble in hydrochloric acid, nor more than 10% of stems and harmless foreign material."

Canadian Government Standard, B.07.025: "**MARJORAM,** whole or ground,

(a) shall be the dried leaves of the marjoram plant;
(b) may contain a small proportion of the flowering tops of the marjoram plant;
(c) shall contain not more than
  (i) 10% stems and foreign material,
  (ii) 16% total ash,
  (iii) 4.5% ash insoluble in hydrochloric acid."

Sweet marjoram contains from about .7 to 3.5% of volatile oil. It is a yellow or greenish-yellow liquid, insoluble in water, soluble in 2 volumes of 80% alcohol, in ether, and in chloroform. Not a great deal is known about the chemistry of this oil. It is reported to consist of about 40% of terpenes, of which terpinene, empirical formula $C_{10}H_{16}$, constitutes the greater part, together with *d*-*α*-terpineol, terpinen-4-ol, and other compounds not yet identified. Terpinene has a lemon-like odor, terpineol a lilac-like odor; these undoubtedly contribute much to the pleasing fragrance of this herb.

V. N. Vashist, M. C. Nigam, K. L. Handa (Regional Res. Lab. Jammu, India) report that the oil distilled from *Majorana hortensis* of European origin grown in Kampur, India, contains a-terpineol, 26.7%; *α*-terpinene, 24.1%; geraniol, 19%; 4-terpinenol, 10.2%; eugenol 8%; and linalool 5.6%.*

Guenther Lossner (Karl-Marx-Univ., Leipzig, Ger.) reports that steam distillation of French leaf marjoram yielded 1.3% (av.) essential oil based on dry material. Thin-layer, column, and gas chromatography of the steam-distilled oil revealed a high proportion of *cis*-sabinene hydrate (25%) besides *trans*-sabinene hydrate (7%). Lossner reports *cis*-sabinene hydrate, when purified by sublimation, as having a strong spicy marjoram-like odor.†

Tarragon

The leaves of tarragon contain a volatile oil, cellulose, pigment, mineral elements, etc.

The chief constituent of the volatile oil of tarragon is reported to be methyl chavicol. Ocimene, phellandrene, and *p*-methoxycinnamaldehyde have

* *Riechstoffe Aromen* 13, 61-2 (1963); *CA.* 58.13. 13705g.
† *Pharmazie* 22(1), 51-4 (1967) (Ger); *CA.* 66. 26. 118799h.

also been reported. Methyl chavicol (estragole, chavicol methyl ether, *p*-allylanisole), empirical formula $C_{10}H_{12}O$, structural formula $CH_2{:}CHCH_2$-$C_6H_4OCH_3$, a phenol ether, is a colorless liquid with an odor slightly reminiscent of anise.

Thyme

Thyme contains a volatile oil, fixed oil, proteins, tannin, cellulose, pentosans, pigment, mineral elements, etc.

United States Government Standards require that the dried leaves of thyme shall contain "not more than 14% of total ash, nor more than 4% of ash insoluble in hydrochloric acid."

Canadian Government Standard, B.07.018: "THYME, whole or ground, shall be the dried leaves and flowering tops of the thyme plant and shall contain not more than

(a) 12% total ash.

(b) 4% ash insoluble in hydrochloric acid."

The volatile oil of thyme is a colorless to yellowish-reddish liquid with a pleasant odor characteristic of the herb, and a sharp taste.

The chief constituent of the oil is thymol (thyme camphor; 3-methyl-6-isopropylphenol), empirical formula $C_{10}H_{14}O$. In the pure state, thymol consists of colorless, translucent crystals possessing an aromatic, thyme-like odor and a pungent taste.

Other organic compounds reported present in the volatile oil of thyme include carvacrol, *l*-linaloöl, *l*-borneol, geraniol, amyl alcohol, β-γ-hexenol, terpinen-4-ol, β-pinene, *p*-cymene, γ-terpinene, caryophyllene, and camphene.

Turmeric

Turmeric contains a volatile oil, fixed oil, resin, a coloring matter, protein, cellulose, pentosans, starch, mineral elements, etc.

Canadian Government Standard, B.07.016: "TURMERIC, whole or ground, shall be the dried rhizome of *Curcuma longa* L."

On distillation, turmeric yields from about 1.3 to 5.5% of volatile oil, orange-yellow in color, slightly fluorescent, and aromatic. Formerly it was said that the oil contained an alcohol, turmerol, and a ketone, curcumone. It is now reported that the main fraction of the volatile oil consists of a monocyclic hydroaromatic ketone, turmerone $C_{15}H_{22}O$, and an aromatic ketone, *ar*-turmerone, or dehydroturmerone, $C_{15}H_{20}O$.*

* Pfau, *Helv. Chim. Acta* 15 (1932), 1482; Rupe, Clar, Pfau, and Plattner, *Helv. Chim. Acta* 17(1934), 380; Ernest Guenther, *The Essential Oils*, vol.2, p.450, Van Nostrand, New York.

Other organic compounds reported present in the oil include *d*-α-phellandrene, *d*-sabinene, zingiberene, cineole, borneol, etc.

The coloring matter of turmeric is curcumin, 1,7-bis (4-hydroxy-3-methoxyphenyl)-1,6-heptadiene-3,5-dione, empirical formula $C_{21}H_{20}O_6$, structural formula $[(CH_3O)(OH)C_6H_3CH{:}CHCO]_2CH_2$. The pure substance is an orange-yellow crystalline powder, insoluble in water, slightly soluble in ether, soluble in alcohol and in glacial acetic acid; it is also soluble in alkalies with a reddish-brown color.

Curcumin assists in the identification of turmeric starch. The yellow, pasty masses of starch become reddish-brown with sodium, potassium, and ammonium hydroxides; they become crimson with concentrated sulfuric acid.

XLIV. pH DETERMINATIONS

The following determinations were made by the author, employing a Beckman pH recording meter. Samples consisted of 2 g of the ground spice contained in 25 cc of distilled water at 25°C.

		pH
Allspice	Jamaica	4.82
	Jamaica, residue following ether extraction	4.82
Allspice	Mexican	5.4
Cassia	Honan	4.32
	Yunan	4.53
	Kwangsi	5.0
	Saigon	4.4
	Batavia	4.6
Cinnamon	Ceylon	4.6
Cloves	Madagascar	4.1
	Zanzibar	4.12
Ginger	Jamaica No. 1	6.2
	Jamaica No. 2	6.08
	African	6.2
Mace		4.6
Nutmeg	East Indies	5.85
	West Indies	5.85
Pepper	black	6.4, 6.42, 6.72, 7.1, 7.18, 7.3, 7.5
Pepper	white	5.2, 6.1, 6.02

No further work was done, and this information is given here for those who may be interested in the hydrogen ion concentration of the spices. It may have some significance, use, and value in checking the purity of ground spices.

Part 3

PHOTOMICROGRAPHS

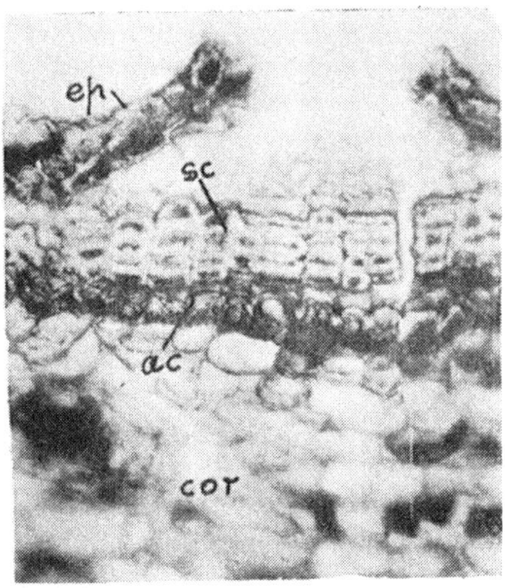

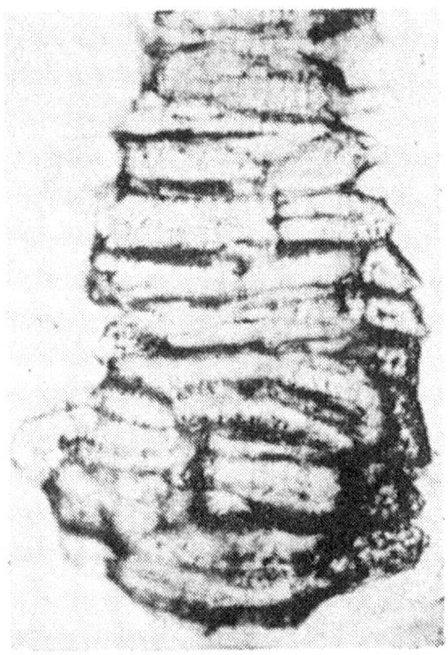

Fig. 41. **Yunan Cassia** (*top*). Phloem fibers (cross section) × 500.

Fig. 42. **Kwangsi Cassia.** (*left*). Stone cork, *sc*; arch cells, *ac* (× 150); epidermis, *ep*; cortex, *cor*.

Fig. 43. **Saigon Cassia** (*right*). Group of stone cells pericycle zone (× 150).

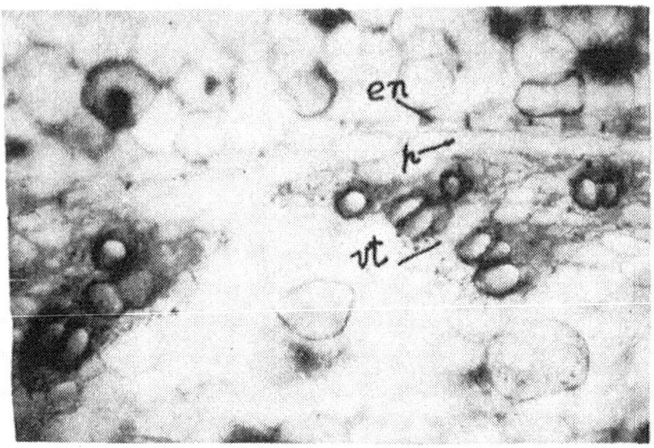

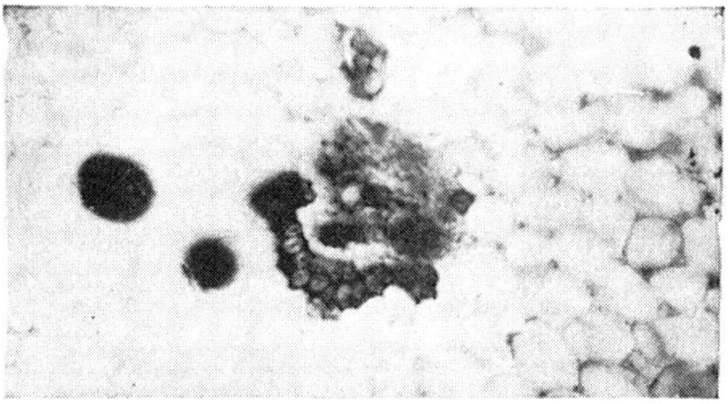

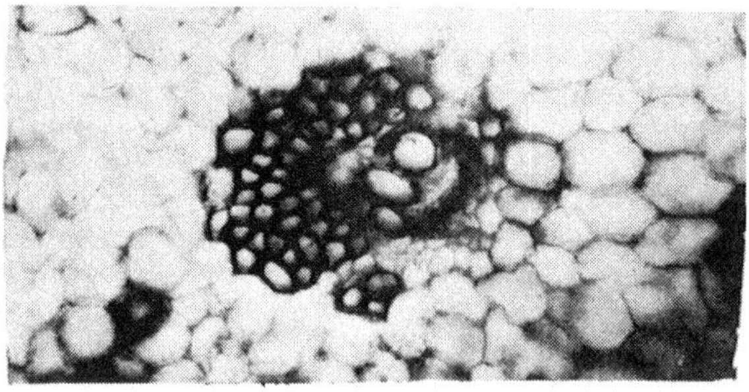

Fig. 44 **Turmeric** × **100** (*top*). Endodermis, *en*; pericycle, *p*; vascular tissue, *vt*.
Fig. 45. **Ginger** × **100** (*middle*). Fibrovascular tissue, cortex.
Fig. 46. **Ginger** × **100** (*bottom*). Fibrovascular tissue, central cylinder.

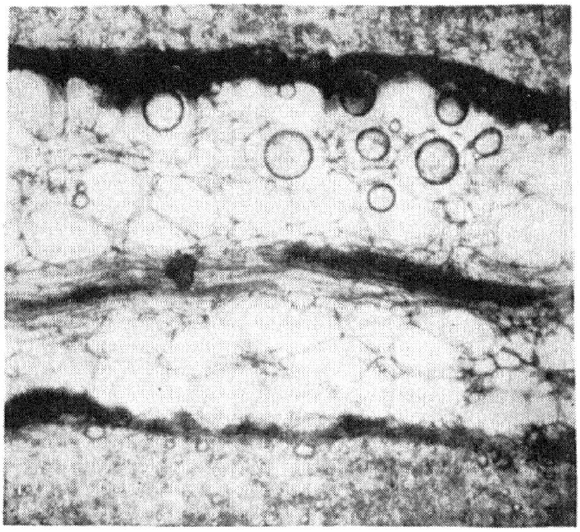

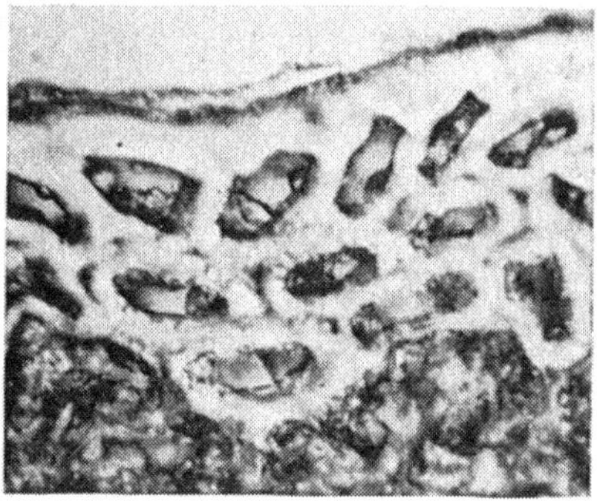

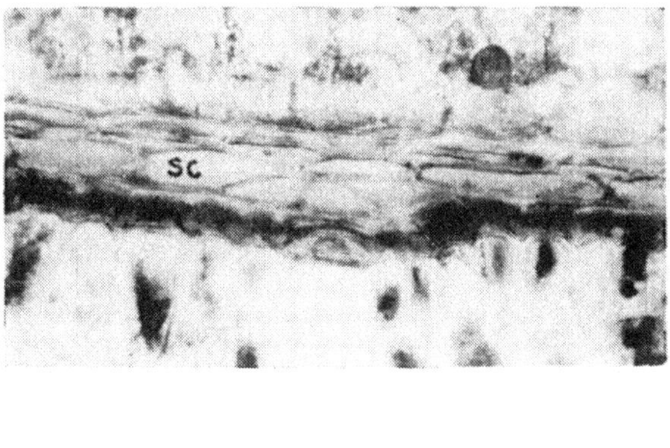

Fig. 47. Nutmeg × **100** (*top*). Oil cells of perisperm.

Fig. 48. Mace, Grenada × **500** (*middle*). Epidermis and hypodermis.

Fig. 49. Pepper × **500** (*bottom*). Seed coat, *sc*; pigment removed.

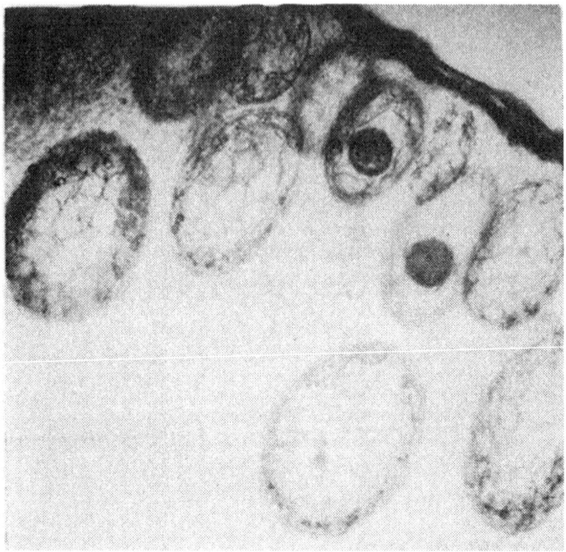

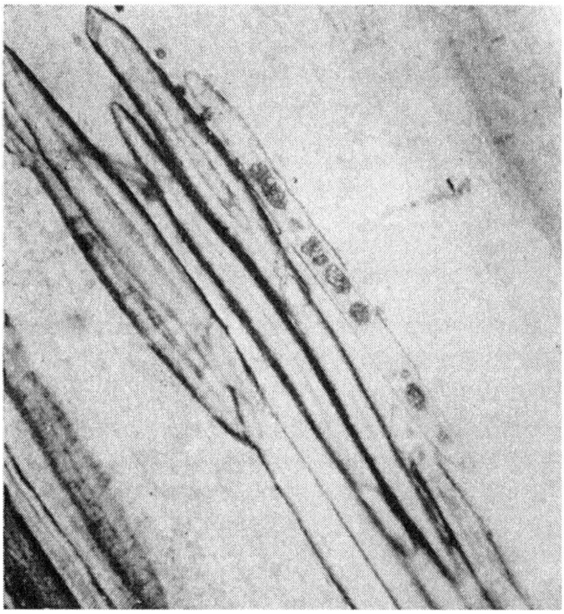

Fig. 50. Clove $\times$ **100** (*top*). Oil cavities.
Fig. 51. Clove $\times$ **150** (*bottom*). Fibers and crystal fiber.

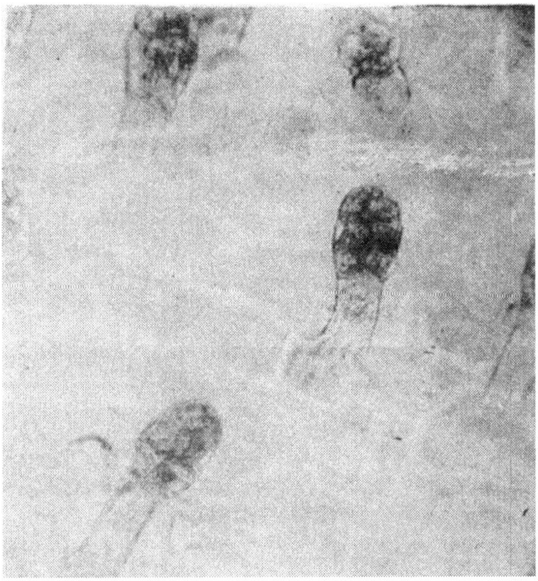

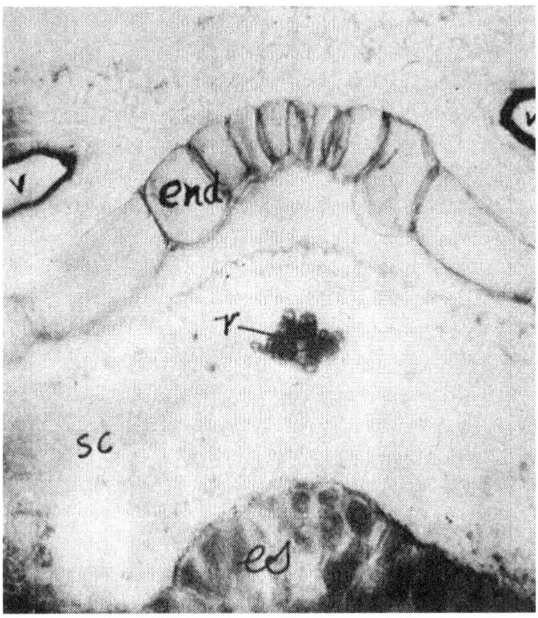

Fig. 52. Chile Ancho × **250** (*top*). Capitate hairs of calyx.

Fig. 53. Caraway × **250** (*bottom*). Commissural side—vittae, *v*; endocarp, *end*; seed coat, *sc*; raphe, *r*; endosperm, *es*.

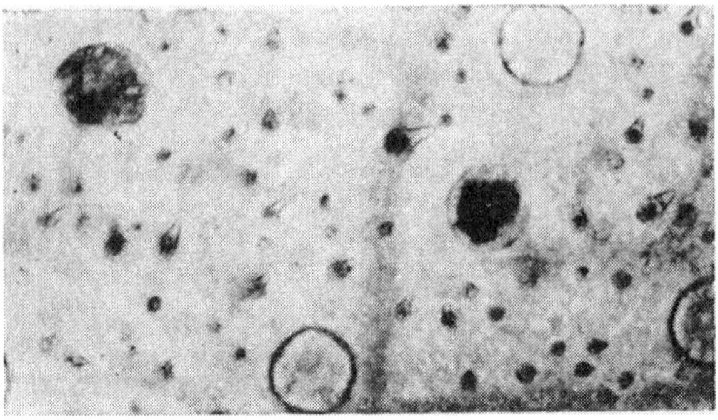

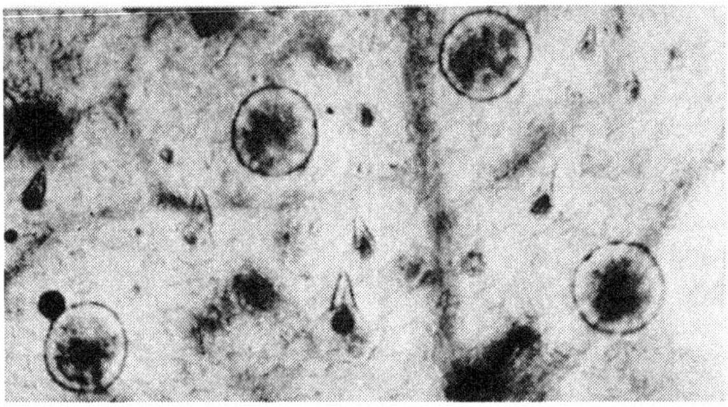

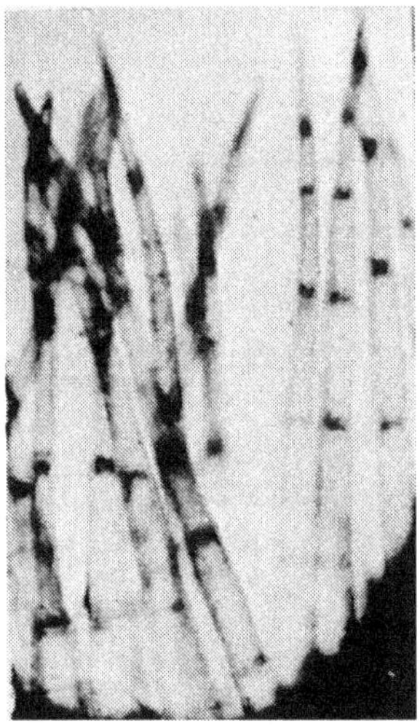

Fig. 54. Savory × **100** (*top*). Upper surface, hairs and oil glands.

Fig. 55. Savory × **100** (*middle*). Lower surface, hairs and oil glands. Note—hairs more numerous on the upper surface.

Fig. 56. Savory × **100** (*bottom*). Throat hairs of calyx.

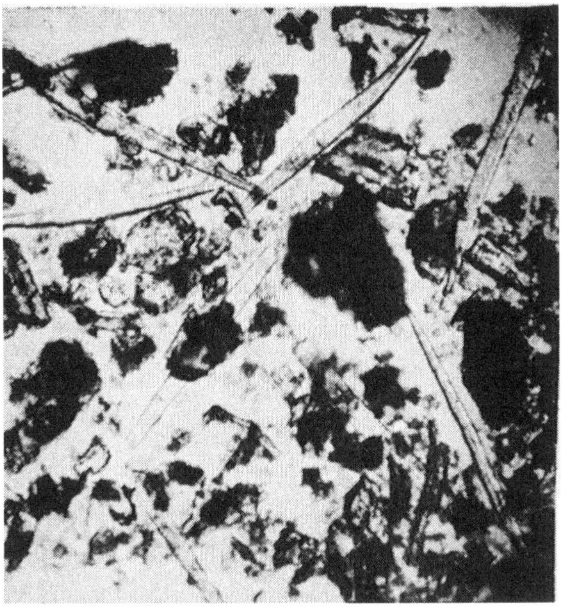

Fig. 57. Ground Ceylon Cinnamon × 100 (*top*). Transmitted light.
Fig. 58. Ground Ceylon Cinnamon × 100 (*botom*). Polarized light.

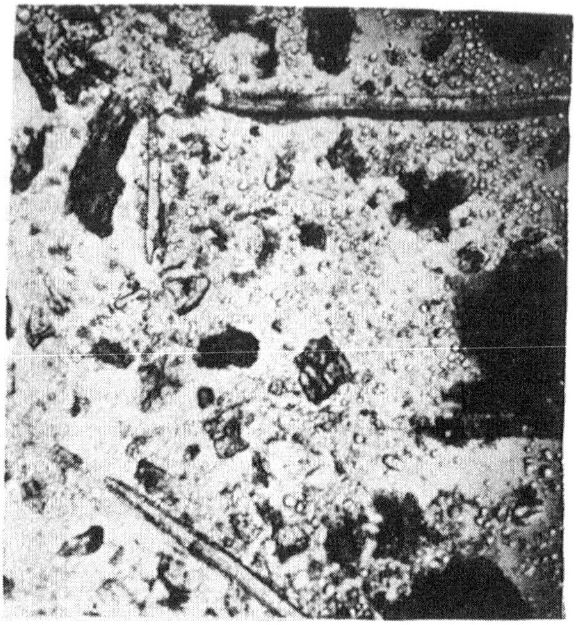

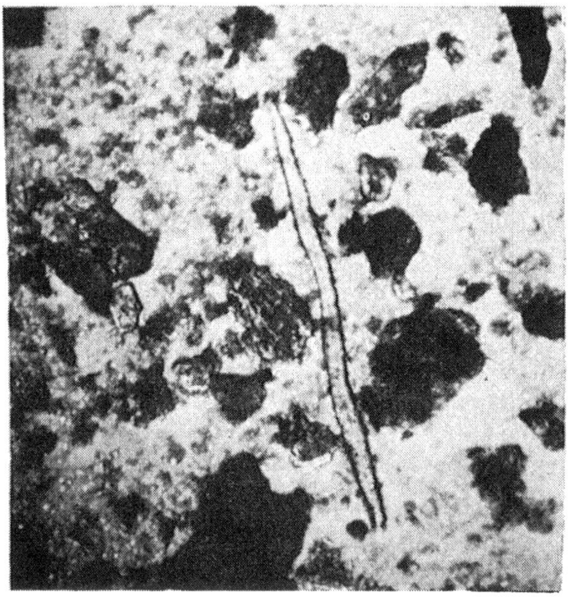

Fig. 59. Ground Batavia Cassia ×**100** (*top*).
Fig. 60. Ground Saigon Cassia ×**100** (*bottom*).

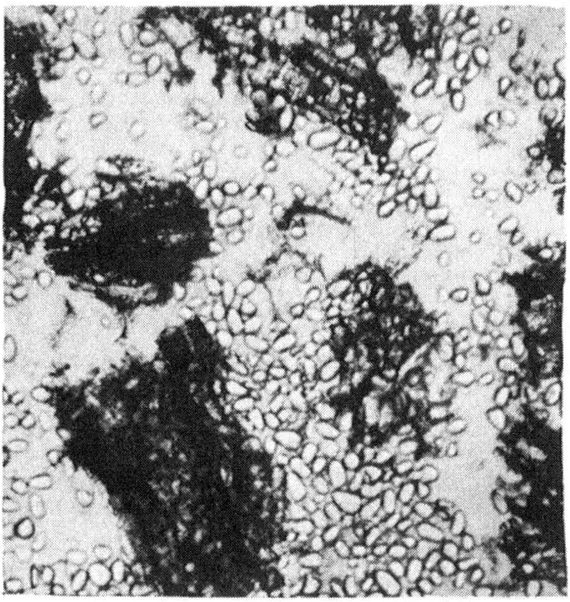

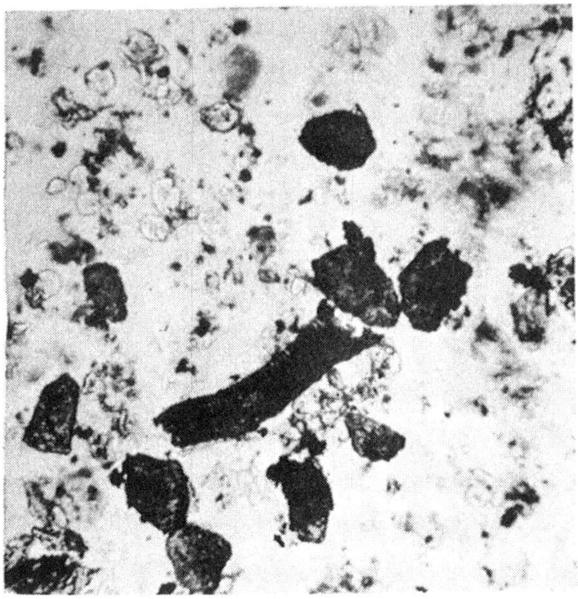

Fig. 61. Ground Ginger ×100 (*top*).
Fig. 62. Ground Turmeric ×100 (*bottom*).

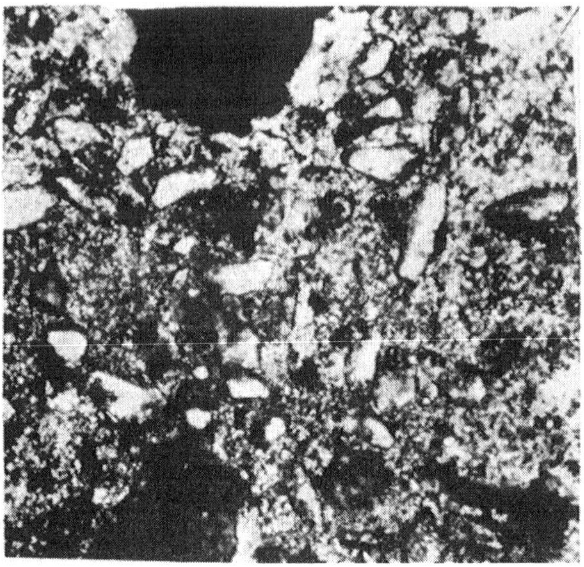

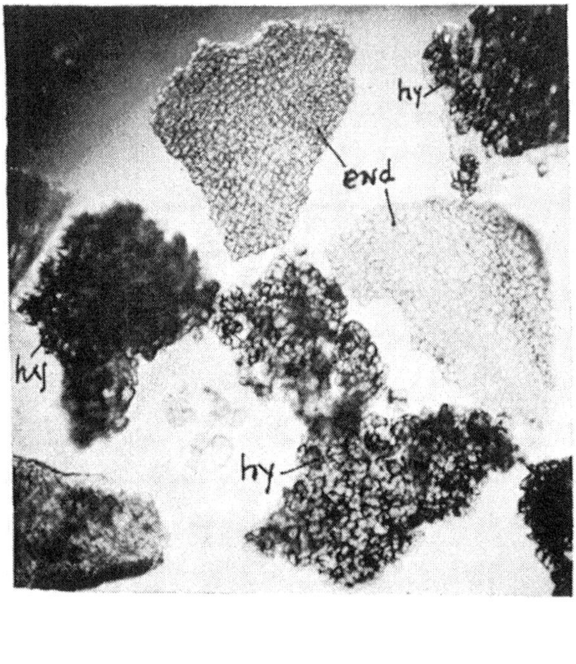

Fig. 63. Ground Black Pepper $\times$ **100** (*top*).
Fig. 64. Black Pepper $\times$ **100** (*bottom*). Fragments of hypodermis, *hy*, and endocarp, *end*.

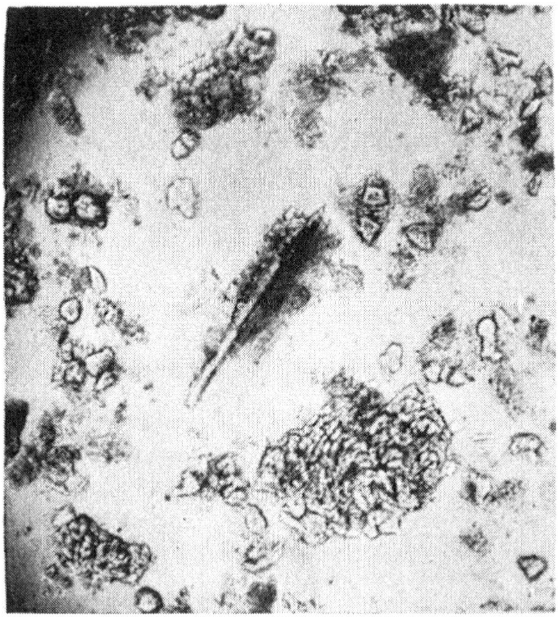

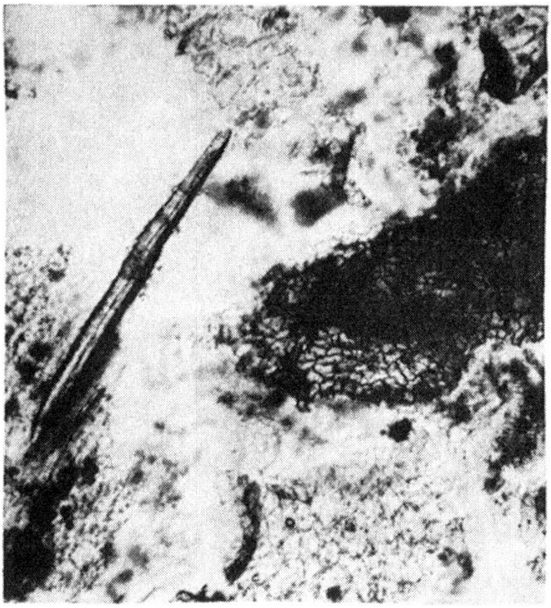

Fig. 65. Ground Allspice $\times$ **50** (*top*).
Fig. 66. Ground Cloves $\times$ **100** (*bottom*).

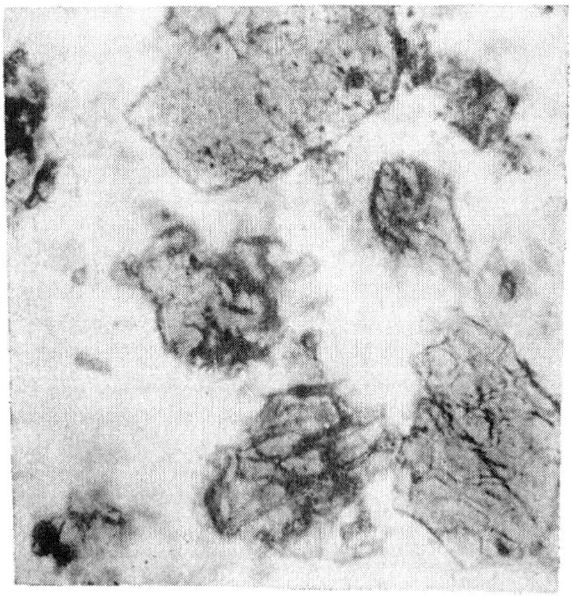

Fig. 67. Ground Cloves $\times$ **100** (*top*).
Fig. 68. Ground Paprika $\times$ **100** (*bottom*).

BIBLIOGRAPHY

Redgrove, Stanley H., *Spices and Condiments*, Sir Isaac Pitman & Sons, Ltd., London, 1933.

Winton, A. L. and K. B., *Structure and Composition of Foods*, Vol. IV, John Wiley & Sons, Inc., New York, 1939.

Guenther, Ernest, *The Essential Oils*, 5 vols., D. Van Nostrand Company, Inc., New York, 1944.

Jacobs, Morris B., *The Chemistry and Technology of Foods and Food Products*, 2 vols., Interscience Publishers, Inc., New York, 1944.

Eckey, E. W., *Vegetable Fats and Oils*, Reinhold Publishing Corporation, New York, 1954.

References

United States Dispensatory, 23rd ed., J. B. Lippincott Company, Philadelphia, 1943.

The Merck Index, 5th ed., Merck & Co., Inc., Rahway, N. J., 1940.

GLOSSARY

Acuminate—Leaf apex tapering to a sharp point.

Acute—Leaf apex broadly coming to a sharp point.

Adnate—United.

Androecium—The stamens collectively.

Annulate—Having rings or ringlike segments.

Aril—A protective covering enveloping the seed.

Axil—The angle formed at the juncture of leaf and stem.

Axillary—Situated in the axil of a leaf.

Bicarpellate—Having two carpels.

Bilabiate—Two-lipped.

Bilocular—Two-celled.

Bilobed—Having two lobes.

Bract—Modified leaf, as of an inflorescence.

Calyx—The sepals collectively form the calyx.

Calyx Tube—The tube of a gamosepalous calyx.

Campanulate—Having the shape of a bell.

Capitate—Shaped like a head, as in gland-tipped hairs.

Carpel—A simple pistil or a unit of a compound pistil.

Collenchyma—Cells with cellulose thickening laid down in longitudinal strips on the walls, chiefly in corners; supporting tissue.

Commissure—The surface by which one carpel joins another as in the *Umbelliferae* fruits.

Commissural—Pertaining to the commissure.

Conduplicate—Folded together lengthwise.

Coriaceous—Leathery in texture.

Corolla—The petals collectively.

Cotyledons—The first leaves of the embryo as formed in the seed.

Cortex—The cellular tissue between the epidermis and the endodermis or the pericycle of roots and stems.

Crenate—Scalloped, or dentate with the teeth much rounded.

Cruciform—Cross-shaped.

Cuticle—A protective layer of waxy substance, chiefly cutin, covering the epidermis.

Cutinized—Deposition of cutin in cell walls; cutin associated with cellulose.

Cyme—A flat-topped inflorescence in which the central flowers bloom first.

Decompound—More than once compound or divided.

Dentate—Having teeth which point outward.

Didynamous—Stamens in two pairs of unequal length.

Dioecious—Unisexual, with staminate flowers on one plant and carpellate on the other.

Discoid—Disk-shaped; round and flat.

Elliptical—Similar to oblong but with continuously rounding sides; like an ellipse.

Endocarp—The inner layer of the pericarp; the inner layer of a fruit.

Endosperm—The cellular tissue in which food is stored in an albuminous seed. In an albuminous seed the food store is outside the embryo. In an exalbuminous seed the food store is contained in the embryo.

Ensiform—Sword-shaped.

Epicarp—The outer layer of the pericarp.

Epigynous—Growing from the top of the ovary.

Exocarp—The outer layer of the pericarp; the outer layer of a fruit.

Filament—The part of the stamen which supports the anther.

Gamopetalous—With united petals; sympetalous.

Gamosepalous—With united sepals; synsepalous.

Glabrous—Without hairs.

Glandular-punctate—Dotted with glands.

Glaucous—Having a whitish coating or bloom.

Gynoecium—A carpel or aggregation of carpels.

Herbaceous—Having the character of a herb; not woody.

Hilum—Scar on a seed, marking the point where the seed broke from the stalk. In starch, the original core around which the layers of starch were laid down.

Hispid—With bristly hairs.

Incised—Cut sharply or irregularly, more or less deeply.

Inferior—Said of an ovary situated below the sepals, petals and stamens.

Imbricate—Overlapping.

Inflorescence—An assemblage of flowers.

Involucel—A secondary involucre.

Involucre—A collection of bracts surrounding a flower cluster or head, or a single flower.

Irregular—Differing in size and shape; shape of cell irregular in form.

Isodiametric—Having approximately the same diameters in all directions.
Labellum—The lip or lower petal.
Labiate—Lipped.
Laciniate—Cut into narrow pointed lobes.
Lanate—Covered with soft hairs; wooly.
Lanceolate—Widening above the base and gradually tapering to a point; lance-shaped.
Leaflet—One of the parts of a compound leaf.
Linear—Long and narrow.
Lobed—Margins cut deeply into lobes or segments.
Locule—The cavity of an ovary in which the ovule or ovules are contained.
Lyrate—With a large terminal lobe and small lower lobes.
Mericarp—One of the carpels of Umbelliferae; half of a schizocarp.
Mesocarp—The middle layer of the pericarp.
Mesophyll—The part of the leaf between the upper and lower epidermis; also in sepals (calyx) and petals (corolla).
Micropyle—The point on the seed at which was the minute opening of the ovule through which the pollen tube passed.
Multiseriate—Cells of hair arranged in several parallel rows or series.
Oblong—Longer than broad and of nearly uniform breadth.
Obovate—Inversely ovate.
Obtuse—With a blunt or rounded apex.
Opposite—Leaf on opposite sides of each node.
Oval—Broadly elliptical; egg-shaped.
Ovary—The part of the pistil in which ovules arise.
Ovate—Shaped like a lengthwise section of an egg.
Palmate—Resembling the open hand.
Panicle—A loose, irregularly branching inflorescence.
Papilionaceous—Having a large posterior petal, the standard, a pair of wings situated at the sides, and a pair of anterior petals, the keel.
Papillose—Bearing minute nipple-like projections; warty.
Parenchyma—Unspecialized, simple plant tissue consisting of thin-walled cells usually of cellulose.
Parietal—Pertaining to the ovary wall.
Pedicel—A flower stalk.
Peduncle—The stalk of a solitary flower or of an inflorescence.
Pericarp—The wall of the ovary when developed or ripened into fruit; the matured ovary wall.
Perisperm—The remaining tissue of the nucleus of the ovary; it is usually stored with food.

Petaloid—Resembling a petal.
Petiolate—Petiole present on leaf.
Petiole—A leaf stalk.
Pinnate—Feather-like.
Pinnatifid—Cleft or parted in a pinnate manner; pinnately cleft.
Pistil—The organ consisting of ovary, style, and stigma.
Placenta—The tissue of an ovary from which ovules arise.
Plumose—Having fine hairs on each side, like the plume of a feather.
Pubescent—Having fine, soft hairs.
Punctate—Dotted with glands.
Pyriform—Pear-shaped.
Raceme—An inflorescence of pediceled flowers upon a common axis.
Racemose—In racemes, or resembling a raceme.
Radicle—The primary root of the embryo as formed in the seed.
Rachis—The axis of a compound leaf.
Receptacle—The part of the floral axis that bears the floral organs.
Reniform—Kidney-shaped.
Reticulated—Having the form of a network.
Revolute—Edges of leaf rolled backward on the lower surface toward the midrib.
Rhizome—An underground root-like stem.
Scalariform—Ladder-like; in the form of a ladder.
Schizocarp—A fruit consisting of two united carpels which separate at maturity into two mericarps; characteristic of the Umbelliferae.
Sclerenchyma—Tissue with cell walls hard, lignified, and usually much thickened.
Segments—Portions of leaf cut to midrib.
Sepal—A leaflike division of the calyx, usually green but sometimes colored.
Serrate—Having teeth which point forward.
Sessile—Without a stalk, as a leaf without a petiole.
Sheathing—Inclosing as by a sheath; leaf invests the stem.
Silique—The seed pod of the *cruciferae* or mustard family.
Sinuate—Having a strongly wavy margin.
Spatulate—Rounded above and gradually tapering down to a narrow base; spoon-shaped.
Spike—A central rachis bearing sessile flowers.
Spinose—Spine-like, or having spines.
Spongy Tissue—Cells of the lower part of the mesophyll with large and numerous intercellular spaces.

Stamen—An organ consisting of a filament and anther, the latter producing pollen grains.
Stigma—The part of the style modified for the reception and germination of pollen grains.
Stone Cells—Sclerenchyma cells varying greatly in size and shape but usually not much longer than they are wide; walls thick, very hard, and strongly lignified.
Style—The part of the pistil between the ovary and the stigma.
Stylopodium—A thickened disk-like expansion at the base of a style, as in *Umbelliferae*.
Sublyrate—Somewhat or slightly lyrate.
Succulent—Juicy, fleshy, having watery tissues.
Suberized—Impregnated with suberin, a waxy or fatty waterproof substance.
Syncarpous—Having united carpels.
Superior—Said of an ovary situated above the sepals, petals, and stamens.
Trifid—Three-cleft.
Trifoliate—Three-leaved.
Truncate—Having the end square, appearing as if cut off.
Tunicate—Having concentric coats, as an onion.
Umbel—A flower cluster in which the flower stalks spring from the same point.
Umbelliferous—Having the mode of flowering called an umbel; a plant which bears umbels.
Undulate—Wavy; rippling.
Venation—Arrangement of veins.
Vittae—Oil secretory canals, as in *Umbelliferae*.
Wooly—Covered with soft hairs; lanate.

INDEX

www.ingramcontent.com/pod-product-compliance
Ingram Content Group UK Ltd.
Pitfield, Milton Keynes, MK11 3LW, UK
UKHW040228290926
15471UKWH00001B/58

9 780820 603780